ASPECTS OF INCOMPLETENESS
Second Edition

LECTURE NOTES IN LOGIC

A Publication of
THE ASSOCIATION FOR SYMBOLIC LOGIC

Editorial Board:
Samuel R. Buss, Managing Editor
Lance Fortnow
Shaughan Lavine
Steffen Lempp
Anand Pillay
W. Hugh Woodin

LECTURE NOTES IN LOGIC   10

ASPECTS OF INCOMPLETENESS
Second Edition

by

Per Lindström
Department of Philosophy
Göteborg University

ASSOCIATION FOR SYMBOLIC LOGIC

A K Peters, Ltd. • Natick, Massachusetts

Addresses of the Editors of Lecture Notes in Logic and a Statement of Editorial Policy may be found at the back of this book.

Sales and Customer Service:
A K Peters, Ltd.
63 South Avenue
Natick, Massachusetts 01760

Association for Symbolic Logic:
C. Ward Henson, Publisher
Mathematics Department
University of Illinois
1409 West Green Street
Urbana, Illinois 61801

Copyright ©1997, 2003 by the Association for Symbolic Logic.

All rights reserved. No part of the material protected by this copyright notice may be reproduced or utilized in any form, electronic or mechanical, including photocopying, recording, or by any information storage and retrieval system, without written permission from the Association for Symbolic Logic.

Library of Congress Cataloging-in-Publication Data

Lindström, Per, 1936-
   Aspects of Incompleteness / by Per Lindström.– 2nd edition.
     p. cm. – (Lecture notes in logic ; 10)
   Includes bibliographical references and index.
   ISBN 1-56881-173-X (pbk. : acid-free paper)
   1. Incompleteness theorems. 2. Recursion theory. I. Title. II. Series.

QA9.65 .L56 2002
511.3–dc21
                                                                                                           2002066232

Publisher's note: This book was published from camera ready copy prepared by the author using the Quark Xpress 5.0 typesetting system. It was printed by IBT Global, New York, on acid-free paper. The cover design is by Richard Hannus, Hannus Design Associates, Boston, Massachusetts.

12 11 10 09 08 07 06 05 04 03   5 4 3 2 1

To Eva
for all she's done for me

PREFACE

By Gödel's incompleteness theorem there is no complete axiomatization of mathematics, not even of first order arithmetic. This leads naturally to the project of investigating the family of, inevitably incomplete, arithmetical theories. In this book we present some of the results obtained in pursuing this aim. A brief summary of the contents of the book is given in the introduction.

Gödel's main idea was to translate metamathematical statements, concerning formulas, formal theories, provability in such theories etc., into arithmetic – the arithmetization of metamathematics. Combining this with the technique of self-reference, he was able to construct a (formal) sentence which is undecidable, neither provable nor disprovable, in a given theory T of arithmetic, thereby demonstrating, in an entirely novel way, that T is incomplete.

The ideas developed by Gödel proved fruitful far beyond their original application. Later essentially new ideas were added. Most important are the basic concepts (and results) of recursion theory – these are needed for a completely general formulation (and proof) of Gödel's theorem – notable are also the so called partial truth-definitions, the idea of formalizing the proof of the completeness of first order logic in arithmetic, and the general concept of interpretability.

In this book we shall be concerned almost exclusively with properties that are common to all sufficiently strong, axiomatizable theories. Thus, for example, investigations of particular theories such as Peano Arithmetic and its fragments, requiring special, model-theoretic or proof-theoretic, methods, fall outside the scope of the book. But even so, the choice of material in some respects reflects the author's preferences; this is particularly true of Chapter 7.

Credits for results, proofs, and exercises, remarks on alternative proofs, related results, etc. are given in the notes following each chapter. Results (exercises) not attributed to anyone are either easy, folklore, or due to the author.

The reader is assumed to be acquainted with (first order) predicate logic including Henkin's completeness proof. We also presuppose knowledge of the elements of recursion theory. Finally, the reader may find our presentation difficult to follow unless (s)he has previously seen a more detailed development of formal arithmetic, an explicitly defined arithmetization of metamathematics, and a full proof of Gödel's theorem.

I am grateful to Daniel Vallström and Rineke Verbrugge; they have very kindly and carefully read substantial portions of this book, pointed out numerous (minor) errors and suggested many improvements. The responsibility for the blemishes that remain is, of course, mine.

May 1997. P. L.

PREFACE TO THE SECOND EDITION

This edition differs from the first edition in several minor and a few major respects. The most important of the latter are the following. There is a new, very simple proof of Theorem 4.5. Similar arguments can be used to prove some related results of Artemov, Beklemishev, and Boolos and these have now been added as Exercises. In Chapter 5 there is a new section (§2) containing some general results on partial conservativity and propositional logic (Theorems 5.7 and 5.8). Finally, Theorem 7.13 has been replaced by a more satisfactory result; the proof, however, is nearly the same. Some Exercises have been omitted, others have been reorganized, and still others have been added.

The bibliography has been expanded and updated.

I am most gratiful to Volodya Shavrukov for much good advice and, particularly, for spotting a false claim in an early version of Chapter 5, §2. He has also, very recently, made an interesting contribution to the subject of this book by proving (in his own words, by putting two and two together) that the first order theories of the lattice of degrees of interpretability and the lattices of Π_n (and Σ_n) sentences modulo provable equivalence in (any given consistent axiomatizable extension of) PA are undecidable (V. Yu. Shavrukov, Effectively dense Boolean algebras in lattices of sentences, in preparation).

September 2002. P. L.

CONTENTS

CHAPTER 0. Introduction 1

CHAPTER 1. Preliminaries 6
  Notes 24

CHAPTER 2. Incompleteness 26
  §1. Incompleteness 26
  §2. Consistency statements 29
  §3. Independent formulas 35
  §4. The length of proofs 38
  Exercises 40
  Notes 48

CHAPTER 3. Numerations of r.e. sets 50
  §1. Numerations of r.e. sets 50
  §2. Types of independence 54
  Exercises 59
  Notes 60

CHAPTER 4. Axiomatizations 62
  §1. Finite and bounded axiomatizability;
      reflection principles 62
  §2. Irredundant axiomatizability 67
  Exercises 69
  Notes 72

CHAPTER 5. Partial conservativity 74
  §1. Partial conservativity 74
  §2. Partial conservativity and propositional logic 83
  Exercises 90
  Notes 94

CHAPTER 6. Interpretability 96
  §1. Interpretability 96
  §2. Faithful interpretability 106
  Exercises 111
  Notes 115

CHAPTER 7. Degrees of interpretability 118
    §1. Algebraic properties 118
    §2. A classification of degrees. 127
    §3. Σ_1 and Π_1 degrees 129
    Exercises 141
    Notes 144

CHAPTER 8. Generalizations 146
    §1. Incompleteness 146
    §2. Axiomatizations. 148
    §3. Interpretability 149
    Notes 152

REFERENCES. 153
INDEX 160
NOTATION 162

0. INTRODUCTION

Let T be a sufficiently strong theory formalized in the language L_A of (first order) arithmetic. Following Gödel, we want to show that there is a sentence φ of L_A which is true (of the natural numbers) but not provable in T. Gödel's idea was to achieve this by constructing φ in such a way that
(*)   φ is true if and only if φ is not provable in T.
Then, assuming (for simplicity) that all theorems of T are true, we are done. For, suppose φ is provable in T. Then, by (*), φ is not true and so, by hypothesis, φ is not provable in T. Thus, φ is not provable in T. But then, by (*), φ is true.

One way to achieve (*) is to find a sentence φ which, in some sense, "says" of itself that it is not provable in T. There are then three major problems that must be solved. First of all, the sentences of L_A deal with natural numbers, they do not deal with syntactical objects such as sentences (of a formal language), proofs, etc. Secondly, even if some of the sentences of L_A can, somehow, be understood as dealing with syntactical objects, it is not clear that it is possible to "say" anything about provability (in T) using only the means of expression available in L_A. And, finally, even if this is possible, there may be no sentence which "says" of *itself* that it isn't provable.

Gödel, however, was able to overcome these difficulties. The first problem is solved by assigning natural numbers to syntactical expressions in a certain systematic way. This is sometimes called a Gödel numbering, and the number assigned to an expression, the Gödel number of that expression. Thus, the numeral of the number assigned to an expression can be regarded as a name of that expression and certain number theoretic statements can be regarded as statements about syntactical objects. (In what follows "φ is a formula", "p is a proof", etc. are short for "φ is the Gödel number of a formula", "p is the Gödel number of a proof", etc.)

To overcome the second difficulty Gödel (re)invented the primitive recursive functions (sets, relations). He showed that a number of crucial properties of (Gödel numbers of) expressions, such as that of being a (well-formed) formula, are primitive recursive. In particular, Gödel showed that, if the set of axioms of T is primitive recursive, this is also true of the relation $PRF_T(\varphi,p)$: p is a proof of the sentence φ in T. φ is provable in T, $PR_T(\varphi)$, if and only if $\exists p PRF_T(\varphi,p)$. This property, however, is not (primitive) recursive.

Gödel then went on to prove that all primitive recursive functions (sets, relations) are definable in L_A. Thus, in particular, there is a formula $Prf_T(x,y)$ of L_A defining $PRF_T(k,m)$. But then $Pr_T(x) := \exists y Prf_T(x,y)$ defines $PR_T(k)$. (In what follows we write $T \vdash \varphi$ for $PR_T(\varphi)$.)

Gödel, however, proved more and this is crucial: for every sentence φ, $T \vdash \varphi$ if and only if $T \vdash Pr_T(\varphi)$. (This is the first time we use the assumption that T is sufficiently strong; but, of course, if T isn't, T is incomplete for that reason.)

This takes care of the second difficulty. So now there remains only the problem of finding a formal sentence which "says" of *itself* that it is not provable in T. Gödel solved this problem in the following way.

Consider the substitution function $SBST(k,m)$ defined by:
$SBST(k,m)=$ the formula obtained from the formula m by replacing the free variable "x" by the numeral of the number k, if m is a formula,
$= 0$ otherwise.

This function is primitive recursive and so is defined in T by a formula $Subst(x,y,z)$ in the sense that for any formula $\delta(x)$ and any number k,
$$T \vdash \forall z(Subst(k,\delta,z) \leftrightarrow z = \delta(k));$$
in other words, for all k, $\delta(x)$, T proves that: "z satisfies $Subst(k,\delta,z)$ if and only if z is the formula $\delta(k)$". Now consider the formula
$$\exists y(Subst(x,x,y) \wedge \neg Pr_T(y)),$$
call it $\gamma(x)$. Let θ be $\gamma(\gamma)$. Intuitively, θ "says" that the result of replacing the variable "x" by the numeral for the number γ in the formula $\gamma(x)$ is not provable in T. But this result, $\gamma(\gamma)$, is θ itself. Thus, θ "says" that θ is not provable in T.

Formalizing this argument we obtain:
(**) $T \vdash \theta \leftrightarrow \neg Pr_T(\theta)$.
(This is an instance of the very important fixed point lemma; Lemma 1, Chapter 1.) And now Gödel's proof can be completed as follows. First we show that
(***) $T \nvdash \theta$.
Suppose $T \vdash \theta$. Then $T \vdash Pr_T(\theta)$. But then, by (**), $T \vdash \neg \theta$ and so T is inconsistent (whence $T \vdash \bot$, where $\bot := \neg 0 = 0$), contrary to assumption.

Thus, (***) holds. But this is exactly what $\neg Pr_T(\theta)$ "says". So $\neg Pr_T(\theta)$ is true and consequently, by (**), θ is true.

Let Con_T be the sentence $\neg Pr_T(\bot)$. Con_T is then a natural formalization of "T is consistent". In proving (***) we actually proved that if $T \vdash \theta$, then $T \vdash \bot$ and so that if T is consistent ($T \nvdash \bot$), then $T \nvdash \theta$. It turns out that this proof can be formalized in T provided that T is sufficiently strong. Thus, $T \vdash \text{Con}_T \to \neg \text{Pr}_T(\theta)$ and so, by (**),

$T \vdash \text{Con}_T \to \theta$.

But then, since $T \nvdash \theta$, it follows that $T \nvdash \text{Con}_T$; in other words, T cannot prove its own consistency. This is Gödel's second incompleteness theorem.

This, in brief, is what Gödel accomplished (restricted to theories in L_A; generalizations to other theories containing arithmetic, for exampel set theory, are straightforward). In Gödel's original proofs it is assumed that the set of axioms is primitive recursive. Subsequently, when the (general) recursive functions had been defined, it turned out, however, that this assumption could, without altering the structure of the proofs, be replaced by the weaker assumption that the set of axioms is recursive. In fact, it became clear that Gödel's first incompleteness theorem holds for all formal systems, in the most general sense, and is actually a result belonging to recursion theory: the set of true (Π_1) sentences of L_A is not recursively enumerable.

In (the above sketch of) Gödel's proof, and in virtually all proofs in the following chapters, the method of arithmetizing metamathematics, i.e., translating metamathematical concepts, statements, etc. into arithmetic, plays a central role. This method is based on a large number of definitions and preliminary results. In Chapter 1 we introduce the basic notation and terminology and state a number of Facts concerning these notions. These Facts will not be proved but references will be given; some of them are proved in almost every exposition of Gödel's theorems, others require quite extensive proofs that would be out of harmony with the rest of the book. In Chapter 1 we also prove the fixed point lemma (Lemma 1.1), the essential undecidability of Robinson's Arithmetic, Q, a very weak finite subtheory of PA, (Theorem 1.2), and the nonexistence of truth-definitions (Theorem 1.3).

In Chapter 2 we present the first and most important results of the subject, Gödel's incompleteness theorem and his (second) theorem on the unprovability of consistency (Theorems 2.1 and 2.4). Gödel's results were subsequently improved in various respects and we present some of these improvements.

The main result of Chapter 3 is that, assuming that T contains a minimum of arithmetic (Q), every recursively enumerable set is numerated by a (Σ_1) formula in T (even if not all Σ_1 sentences provable in T are true)

(Theorem 3.1). We also prove some refinements of this result.

Given that the set of axioms of a theory T is infinite, it is natural to ask if these axioms can be replaced by a finite set of axioms. In Chapter 4 we apply the so called reflection principles to prove some negative results concerning this problem (Theorems 4.1, 4.2). For example, neither Peano Arithmetic, PA, nor any one of its consistent extensions (in L_A) is finitely axiomatizable. On the other hand, every extension of PA has an axiomatization which is irredundant in the sense that none of the axioms can be derived from the other axioms (Theorem 4.6). We also prove the existence of not irredundantly axiomatizable theories (Theorem 4.7).

Let Γ be a set of sentences, for example Σ_n or Π_n. A sentence φ is Γ-conservative over T if every sentence in Γ provable in T + φ is provable already in T. Partial conservativity is studied in Chapter 5, where the basic existence theorems are proved (Theorems 5.2, 5.3, 5.4); it plays an important role in Chapters 6 and 7.

An interpretation of a theory S in a theory T is, roughly speaking, a function t on the set of formulas of S into the set of formulas of T such that t preserves logical form and $T \vdash t(\varphi)$ whenever $S \vdash \varphi$. S is interpretable in T, $S \leqslant T$, if there is an interpretation of S in T. If, in addition, $S \vdash \varphi$ whenever $T \vdash t(\varphi)$, we shall say that t is faithful and that S is faithfully interpretable in T.

Interpretability was originally used as a tool in proofs of (relative) consistency and undecidability. Interpretability (in arithmetical theories) is studied for its own sake in Chapter 6. The key result is that if T is an extension of PA and Con_S is a sentence which (in a suitable sense) in T "says" that S is consistent, then $S \leqslant T + Con_S$ (the arithmetization of Gödel's completeness theorem; Theorem 6.4). From this it follows that $S \leqslant T$ if and only if for every finite subtheory S' of S, $T \vdash Con_{S'}$ (Lemma 6.2) and that if S, too, is an extension of PA, then $S \leqslant T$ if and only if every Π_1 sentence provable in S is provable in T (Theorem 6.6). We also prove similar characterizations of faithful interpretability (Theorems 6.13, 6.14).

Mutual interpretability is an equivalence relation; its equivalence classes will be called degrees of interpretability. Let T be a consistent extension of PA. The degrees of extensions of T, partially ordered by the relation induced by $\leqslant$, form a distributive lattice (Theorem 7.1). This lattice is studied in Chapter 7 both from a purely algebraic point of view and in terms of the nature of the theories belonging to a given degree.

It is quite often true in the following pages that a result stated for extensions of PA actually holds for all extensions of some (much) weaker, sometimes finitely axiomatizable, subtheory of PA. We shall, however, pay little attention to facts of this type: what we are mainly interested in here are the properties shared by all theories containing a sufficient amount of arithmetic. But, if a result is (proved to be) true of Q (and its extensions), this will be explicitly noted.

Almost all the results presented in this book hold in a very general setting. In spite of this we shall in Chapters 1–7, for reasons of simplicity and readability, formulate (and prove) these results for theories formalized in L_A only. We partly make up for this lack of generality in Chapter 8, which is devoted to generalizations, usually straightforward, to theories formalized in other languages, the most important being the language of (first-order) set theory.

1. PRELIMINARIES

In this chapter we introduce the basic notation and terminology which will be used throughout this book. We also state a number of basic Facts. These Facts will not be proved; some of them are rather obvious (and easy to believe), others are substantial and well-known theorems; further Facts will be stated when they are needed. These Facts are sufficient for most of the proofs in this book; the chief exceptions are the proofs of Theorems 3.5, 5.7, and 6.4. Finally, we prove the very important Fixed Point Lemma (Lemma 1) and apply it to prove that Robinson's Arithmetic Q is essentially undecidable (Theorem 2) and that in extensions of Q there are no truth-definitions (Theorem 3).

The language L_A of elementary arithmetic can be described as follows. The alphabet consists of:

the propositional connectives: $\neg, \wedge, \vee, \rightarrow, \leftrightarrow$,
the quantifiers: $\exists, \forall$,
the equality symbol: $=$,
symbols used to form (individual) variables: v, ',
parentheses: (,),
the arithmetical constants: $0, S, +, \times$.

(The intended interpretation of S is the successor function.) Thus, the alphabet is finite. The *variables* of L_A are the expressions v, v', v'', etc. We write v_n for v followed by n occurrences of '. In most contexts x, y, z, u, v, w, possibly with subscripts etc., will be used for variables. The terms, formulas, and sentences of L_A are defined as usual. Among the terms we distinguish the *numerals* 0, S0, SS0, SSS0, ... These will be written 0, 1, 2, ... Thus, we shall omit bars and other devices ordinarily used to indicate numerals (or Gödel numbers) and use the same symbols for natural numbers and for formal numerals. In most cases this will cause no trouble as long as the symbols for formal variables are kept strictly apart from the symbols for natural numbers (and numerals). For the latter we use k, m, n, p, q, r, s, possibly with subscripts etc. and symbols for formulas (see below).

For sentences and formulas of L_A we use lower case Greek letters. Sentences will be written as $\varphi, \psi, \theta, \chi$, etc. and formulas as $\alpha(x), \beta(x,y), \gamma(x), \xi(x), \eta(x_1,...,x_n), \rho(x,x'), \tau(x), \xi, \gamma$, etc. The variables displayed are almost always exactly the free variables of the formula. $\xi(y)$ is obtained from $\xi(x)$ by replacing x by y, assumed not to be free in $\xi(x)$, and, possibly renaming bound variables in the usual way. $\xi(k)$ is obtained from $\xi(x)$ by replacing x by the numeral k (or, if you prefer, by the numeral for the number k). This generalizes in the obvious way to substitutions involving more than one variable. We use := to denote equality between formulas. Let $\top := 0 = 0$ and $\bot := \neg \top$.

By a *theory* T we understand a set of sentences (to be thought of as the (nonlogical) axioms of T). (It would be inconvenient to identify a theory T with the set of its theorems, since quite often we need to know that there is a formula binumerating (defined below) (the set of axioms of) T.) Note that, although we shall mainly be interested in theories that are reasonable from an arithmetical point of view, such reasonableness is not part of the concept *theory*. $T + \varphi$ is the theory obtained from T by adding φ as a (new) axiom. $T + X$, where X is a set of sentences, is understood similarly. We assume given a fixed complete deductive calculus PL for first order logic. Referring to PL certain (finite) formal objects (sequences of sentences) are *proofs* (in T). A proof is a *proof of* its last sentence. The sentence φ is *provable in* T, $T \vdash \varphi$, if there is a proof of φ in T. $T \vdash \xi(x_1,\ldots,x_n)$, where $x_1,\ldots,x_n$ are all the free variables of $\xi(x_1,\ldots,x_n)$, is short for $T \vdash \forall x_1 \ldots x_n \xi(x_1,\ldots,x_n)$. Th(T) is the set of *theorems* of, i.e., sentences provable in, T. If X is a set of sentences, we write $T \vdash X$ or $X \dashv T$ to mean that $T \vdash \varphi$ for every $\varphi \in X$. Thus, $S \dashv T$ means that S is a *subtheory* of T (T is an *extension* of S). We write $\vdash \varphi$ for $\emptyset \vdash \varphi$, where $\emptyset$ is the empty set. Thus, $\vdash \varphi$ means that φ is provable in logic (PL).

Let N be the set of natural numbers. $\mathbf{N} = (N, +, \times, S, 0)$ is the standard model of arithmetic. A sentence φ is *true* if it is true in $\mathbf{N}$. A theory is *true* if all its axioms (and therefore, all its theorems) are true.

There are two (true) theories PA (Peano Arithmetic) and Q (Robinson's Arithmetic) that will play a prominent role in what follows. Q is a finite theory; its axioms are (we omit the initial universal quantifiers):

Q1 $\quad Sx = Sy \to x = y$,
Q2 $\quad \neg 0 = Sx$,
Q3 $\quad \neg 0 = x \to \exists y(x = Sy)$,
Q4 $\quad x + 0 = x$,
Q5 $\quad x + Sy = S(x + y)$,
Q6 $\quad x \times 0 = 0$,
Q7 $\quad x \times Sy = (x \times y) + x$.

We introduce the two-place predicates $\leq$ and $<$ by means of the definitions:

$\quad x \leq y =_{df} \exists z(z + x = y)$,
$\quad x < y =_{df} x \leq y \land \neg x = y$.

With our present simplified notation certain (harmless) ambiguities arise. For example, 2 + 3 can be read as a numeral but also as an expression containing the symbol +. In Fact 1 below we have indicated that the latter is the intended reading by underscoring the relevant function symbol. But, of course, terms such as Sy, x + y, x + 3, 4 × z, etc. are unambiguous. Also, with very few exceptions, it is, in view of Fact 1, not important which way, say, 2 + 3 is understood.

Fact 1. The following formulas are provable in Q for all k, m,
(i) $\neg k = m$ for $k \neq m$,
(ii) $k \underline{+} m = k + m$,
(iii) $k \underline{\times} m = k \times m$,
(iv) $x \leq m \to x = 0 \vee x = 1 \vee \ldots \vee x = m$,
(v) $x \leq m \vee m \leq x$.

Q is a very weak theory. The sentences $\forall x(0 + x = x)$, $\forall x \neg(x = Sx)$, cannot be proved in Q.

The axioms of PA consist of the axioms of Q plus the (universal closures) of formulas of the form
$$\alpha(0) \wedge \forall x(\alpha(x) \to \alpha(Sx)) \to \forall x \alpha(x).$$
(Here $\alpha(x)$ may contain free variables other than x.) This is the *induction scheme* and is as close as we can get to the full (second order) induction axiom in first order arithmetic.

From the induction scheme we can derive the *least number principle*: for every formula $\alpha(x)$ as above,
$$\text{PA} \vdash \exists x \alpha(x) \to \exists x(\alpha(x) \wedge \forall y(y < x \to \neg \alpha(y))).$$
Obviously, $Q \dashv \text{PA}$. In PA axiom Q3 is redundant. $\forall xy(x \leq y \vee y \leq x)$ is provable in PA, but not in Q. In fact, this is sometimes the sole reason for writing "PA ⊢" rather than "Q ⊢".

Gödel proved that every primitive recursive function is definable in first order arithmetic. Formalizing this proof, he proved that:

Fact 2. For each primitive recursive function $f(k_0, \ldots, k_n)$ there is a formula $\delta_f(x_0, \ldots, x_n, y)$ such that for all $k_0, \ldots, k_n$,
(i) $Q \vdash \delta_f(k_0, \ldots, k_n, y) \leftrightarrow y = f(k_0, \ldots, k_n)$,
(ii) $\text{PA} \vdash \delta_f(x_0, \ldots, x_n, y) \wedge \delta_f(x_0, \ldots, x_n, z) \to y = z$,
(iii) $\text{PA} \vdash \exists y \delta_f(x_0, \ldots, x_n, y)$.

In Fact 2(i) $f(k_0, \ldots, k_n)$ is, of course, a numeral, i. e. does not contain the symbol f.

A formula $\delta(x_0,\ldots,x_n,y)$ such that for all $k_0,\ldots,k_n$,
$$T \vdash \delta(k_0,\ldots,k_n,y) \leftrightarrow y = f(k_0,\ldots,k_n)$$
will be said to *define* f *in* T.

For (general) recursive functions we have the following weaker Fact (for Fact 3(b), see below):

Fact 3. (a) For every (total) recursive function $f(k_0,\ldots,k_n)$, there is a formula $\delta_f(x_0,\ldots,x_n,y)$ defining f in Q.

The formula $\rho(x_0,\ldots,x_n)$ *numerates* the relation $R(k_0,\ldots,k_n)$ *in the theory* S if for all $k_0,\ldots, k_n$ (as usual "iff" is short for "if and only if"),
$$R(k_0,\ldots,k_n) \text{ iff } S \vdash \rho(k_0,\ldots,k_n).$$
Thus, $\xi(x)$ numerates X in S if for every k,
$$k \in X \text{ iff } S \vdash \xi(k).$$
$\rho(x_0,\ldots,x_n)$ *binumerates* the relation $R(k_0,\ldots,k_n)$ *in* S if for all $k_0,\ldots,k_n$,
$$R(k_0,\ldots,k_n) \text{ iff } S \vdash \rho(k_0,\ldots,k_n),$$
$$\text{not } R(k_0,\ldots,k_n) \text{ iff } S \vdash \neg\rho(k_0,\ldots,k_n).$$
In particular, $\xi(x)$ binumerates X in S if for every k,
$$k \in X \text{ iff } S \vdash \xi(k),$$
$$k \notin X \text{ iff } S \vdash \neg\xi(k).$$
If a formula binumerates X (R) in S, it binumerates X (R) in every consistent extension of S.

If S is recursively enumerable (r.e.), any set (relation) numerated by some formula in S is r.e. and any set (relation) binumerated by some formula in S is recursive.

Fact 3(a) has the following:

Corollary 1. (a) A set X (relation R) is recursive iff there is a formula binumerating X (R) in Q.

(b) A set X (relation R) is r.e. iff there is a formula numerating X (R) in Q.

This corollary and most of those below in this chapter are easy consequences of the relevant Facts; their proofs are, therefore, left to the reader.

Note that, in view of Corollary 1, we have the remarkable fact any set X (relation R) which is (bi)numerated by some formula in some r.e. theory, is (bi)numerated by a (possibly different) formula already in Q.

We write $\exists x \leqslant y \beta(x)$ for $\exists x(x \leqslant y \wedge \beta(x))$ and $\forall x \leqslant y \beta(x)$ for $\forall x(x \leqslant y \rightarrow \beta(x))$. $\exists x < y \beta(x)$ and $\forall x < y \beta(x)$ are defined in a similar way. The initial quantifiers of these formulas are *bounded.*

A formula is *primitive recursive in the strict sense* (*SPR*) if it is of the form $\delta_f(x_0,\ldots,x_n,0)$, where f is primitive recursive and $\delta_f(x_0,\ldots,x_n,y)$ is as in Fact 2. We define the *primitive recursive* (*PR*) formulas to be the members of the least set F of formulas containing the SPR formulas such that F is closed under propositional connectives, bounded quantification, replacing variables by numerals, and if ξ is a member of F and $\delta_f(x_0,\ldots,x_n,y)$ is as in Fact 2 with f primitive recursive, then $\exists z(\delta_f(x_0,\ldots,x_n,z) \wedge \xi)$ and $\forall z(\delta_f(x_0,\ldots,x_n,z) \rightarrow \xi)$ are members of F. (Every PR formula is, provably in PA, equivalent to an SPR formula.)

Exactly which formulas turn out to be PR will depend on the details of the proof of Fact 2. However, regardless of those details we have the following consequence of Fact 2. A formula $\eta(x)$ is *decidable* in T if for every k, $T \vdash \eta(k)$ or $T \vdash \neg \eta(k)$; and similarly for formulas with more than one free varaible; a sentence φ is *decidable* in T if either $T \vdash \varphi$ or $T \vdash \neg \varphi$.

Corollary 2. If $\rho(x_1,\ldots,x_n)$ is PR, then $Q \vdash \rho(k_1,\ldots,k_n)$ iff $\rho(k_1,\ldots,k_n)$ is true. It follows that
(i) every PR formula is decidable in Q,
(ii) a set X (relation R) is primitive recursive iff there is a PR formula binumerating X (R) in Q.

In what follows Corollary 2 will be applied without further mention.

A PR formula binumerating X (R) in Q will be called a PR *binumeration of* X (R). A *numeration of* X is a formula numerating X in PA.

Suppose PA $\dashv$ T. Then, by Fact 2, if $f(k_0,\ldots,k_n)$ is a primitive recursive function and $\delta_f(x_0,\ldots,x_n,y)$ is the corresponding formula, we can add the function symbol f to the language of T and add
$$f(x_0,\ldots,x_n) = y \leftrightarrow \delta_f(x_0,\ldots,x_n,y)$$
as a new axiom. (Thus, we shall be using the same function-symbol in the object language as in the metalanguage.) The resulting theory S is then a conservative extension of T in the sense that every sentence in the language of T provable in S is provable already in T. Thus, we may assume that f is a symbol in the language of T. Occasionally, the choice of the defining formula $\delta_f(x_0,\ldots,x_n,y)$ is essential, e.g. in the proofs of Theorems 3.5 and 5.7, but most of the time it is not.

In particular, we shall use the function symbols $\langle x,y \rangle$ and $(x)_y$ for the primitive recursive functions $\langle k,m \rangle$ and $(k)_m$ defined by:
$$\langle k,m \rangle = 2^k \times 3^m,$$

$(k)_m$ = the number n such that p_m^n divides k, but p_m^{n+1} doesn't if k > 0,
= 0 if k = 0.

(Here p_m is the mth prime number: $p_0 = 2$, $p_1 = 3$, etc.) The function $(k)_m$ will be used to code finite sequences of natural numbers; namely, for each finite sequence $n_0,..., n_k$ of natural numbers, there is a number n such that $(n)_i = n_i$ for $i \leq k$.

The function $(k)_m$ can be used to transform an inductive definition into an explicit definition in PA in the following way. Suppose, for example, f(k) is defined by:

f(0) = 0,
f(n+1) = g(f(n),n) if n ∈ X,
= h(f(n)) if n ∉ X.

Suppose g(k,m), h(k), and X are formally represented by g, h, and $\xi(x)$. Let $\delta(x,y) :=$

$\exists z((z)_0 = 0 \wedge \forall u < x(\xi((z)_u) \to (z)_{u+1} = g((z)_u, u)) \wedge$
$\forall u < x(\neg \xi((z)_u) \to (z)_{u+1} = h((z)_u)) \wedge (z)_x = y)$.

Then $\delta(x,y)$ defines f in PA and

PA $\vdash \forall x \exists y \forall z (\delta(x,z) \leftrightarrow z = y)$.

Thus, we may introduce a function symbol f by means of the definition:

$f(x) = y \leftrightarrow \delta(x,y)$.

It is then easy to see that the formalizations of the clauses of the definition of f(k) become provable in PA:

PA $\vdash f(0) = 0$,
PA $\vdash \xi(x) \to f(x+1) = g(f(x),x)$,
PA $\vdash \neg \xi(x) \to f(x+1) = h(f(x))$.

We assume given a Gödel numbering of the formal objects of L_A (extensions of L_A obtained by adding symbols for certain functions) among them all proofs. Since there is really no reason to distinguish between a formal object and its Gödel number, we shall "identify" the two. (We do not really care exactly what the formulas, proofs etc. of a theory are; the only thing that matters is how they are related to one another.) Thus, on the pages of this book you will find no formulas of L_A, only "formulas" referring to such formulas. (But the reader may, of course, still think of formal objects as strings of symbols.) We shall assume, as we clearly may, that if p is a proof (in some theory T) of φ, then $\varphi < p$.

Formulas and sentences being numbers, it follows that symbols for formulas and sentences are (symbols for) numerals. Thus, for example, $\xi(\eta(y))$ makes perfectly good sense; it is the result of replacing x in $\xi(x)$ by (the numeral for) $\eta(y)$. (Note that y is not free in $\xi(\eta(y))$.) $\xi(\eta(k))$ is obtained by first replacing y by k in $\eta(y)$, giving $\eta(k)$, and then replacing x by $\eta(k)$ in $\xi(x)$.

The Gödel numbering can be defined in such a way that everything, that should be primitive recursive, is. In particular, the following is true (see also Fact 4(d) below):

Fact 4. (a) The function corresponding to concatenation is primitive recursive.

(b) The function corresponding to substitution of numerals for variables is primitive recursive.

(c) The sets of formulas and sentences are primitive recursive.

By Fact 4(a), $\neg\varphi$, $\varphi \to \psi$, etc. are primitive recursive functions of φ and ψ and so we may, and shall, use $\neg$, $\to$, etc. as formal symbols for these functions and write $\neg x$, $x \to y$, etc. for $\neg(x)$, $\to(x,y)$, etc.

As has already been mentioned, in many cases our (simplified) notation is not unambiguous. For example, $x = \varphi \to \psi$ can be read in three different ways. One of these is eliminated by writing $x = (\varphi \to \psi)$. But this formula is still ambiguous: does it contain the function symbol $\to$, or doesn't it? The answer to this and similar questions will always be clear from the context, when it matters. For example, we are allowed to add the symbol f for an (arbitrary) primitive recursive function f to the vocabulary of T only if we have assumed that PA $\dashv$ T, and then it doesn't really matter which way $f(2+3)$, say, or $\varphi \to \psi$, occurring as a term, is understood. On the other hand, terms such as $f(x)$ or $y \to \varphi$ are, of course, unambiguous.

In this book we shall be interested in r. e. theories only. In most contexts it is not necessary to distinguish between deductively equivalent theories. Thus, we may take advantage of the following result known as Craig's theorem.

Theorem 1. For any r. e. set X, there is a primitive recursive set Y such that $Y \dashv \vdash X$.

Proof. If $X = \emptyset$, this is trivial. Suppose $X \neq \emptyset$. There is a primitive recursive function f such that $X = \{f(k) : k \in \mathbb{N}\}$. For any sentence φ, let $\varphi^{(0)} := \varphi$ and $\varphi^{(n+1)} := \varphi^{(n)} \wedge \varphi$. Let $Y = \{f(k)^{(k)} : k \in \mathbb{N}\}$. ∎

In view of Craig's theorem we may adopt the first of the following three conventions; the other two are introduced to avoid needless repetition:

Convention 1. All theories denoted by single (decorated) letters, S, S_0, T, T', A, B etc. are primitive recursive.

Convention 2. All theories denoted by single (decorated) letters, S, S_0, T. T', A, B, etc. are consistent.

Convention 3. From now on until Chapter 8, T is an extension of Q, $Q \dashv T$. If a theory is written as S, S', S_0, etc. this is meant to indicate that, unless the contrary is explicitly assumed, the fact that this theory is formalized in L_A is really irrelevant.

We now define the *arithmetical hierarchy* of formulas (sentences) of L_A, in other words, the sets Σ_n and Π_n in the following way. Σ_n and Π_n are the least sets containing PR, closed under $\wedge$, $\vee$, and bounded quantification and such that (i) $\Sigma_n \cup \Pi_n \subseteq \Sigma_{n+1} \cap \Pi_{n+1}$, (ii) if ξ is Σ_n (Π_n), then $\neg \xi$ is Π_n (Σ_n), (iii) if ξ_0 is Σ_n (Π_n) and ξ_1 is Π_n (Σ_n), then $\xi_0 \to \xi_1$ is Π_n (Σ_n), (iv) if ξ is Σ_n (Π_n) and $\delta_f(x_0,\ldots,x_n,y)$ is as in Fact 2, then $\exists z(\delta_f(x_0,\ldots,x_n,z) \wedge \xi)$ and $\forall z(\delta_f(x_0,\ldots,x_n,z) \to \xi)$ are Σ_n (Π_n), (v) Σ_n is closed under existential quantification, and (vi) Π_n is closed under universal quantification. It follows that $\Sigma_0 = \Pi_0 =$ PR. B_n is the set of Boolean combinations of Σ_n formulas. Let Φ be either Σ_n or Π_n or B_n. Then $\Phi^T = \{\xi \colon \exists \eta \in \Phi \colon T \vdash \xi \leftrightarrow \eta\}$. A formula is Δ_n^T if it is Π_n^T and Σ_n^T or Σ_n and Π_n^T; $\Delta_n = \Delta_n^{PA}$.

In what follows Γ is either Σ_{n+1} or Π_{n+1} and Γ^+ is either Σ_n or Π_n. Γ^d, the *dual* of Γ, is Σ_n, if Γ is Π_n, and Π_n, if Γ is Σ_n. *In writing* Σ_n, Π_n, Δ_n *or* B_n *we almost always omit the (obvious) assumption that* $n > 0$.

The arithmetical hierarchy generalizes to formulas containing new symbols for primitive recursive functions in the obvious way: if $\xi(x)$ is Γ^+ and $g(x_0,\ldots,x_n)$ is primitive recursive, then $\xi(g(x_0,\ldots,x_n))$ is Γ^+. In particular, Γ^+ is closed under $\forall x \leqslant f(x_0,\ldots,x_n)$ and $\exists x \leqslant f(x_0,\ldots,x_n)$.

From the definition of Σ_n and Π_n and Fact 4(a), (b), (c) we get:

Fact 4. (d) The sets Σ_n and Π_n are primitive recursive.

Fact 5. (a) For each Σ_{n+1} formula (sentence) σ, we can effectively find a Π_n formula $\pi(x)$ such that PA $\vdash \sigma \leftrightarrow \exists x \pi(x)$.

(b) For each Π_{n+1} formula (sentence) π, we can effectively find a Σ_n formula $\sigma(x)$ such that PA $\vdash \pi \leftrightarrow \forall x \sigma(x)$.

By Fact 5, if we are working in an extension of PA, we can always assume that any Σ_{n+1} formula (sentence) is of the form $\exists x \pi(x)$, where $\pi(x)$ is Π_n and that any Π_{n+1} formula (sentence) is of the form $\forall x \sigma(x)$, where $\sigma(x)$ is

Σ_n. Also note that it follows from Fact 5 that for each Σ_{n+1} formula $\sigma(x_0,\ldots,x_{k-1})$ there is a PR formula $\rho(x_0,\ldots,x_{k-1},y_0,\ldots,y_n)$ such that
$$PA \vdash \sigma(x_0,\ldots,x_{k-1}) \leftrightarrow \exists y_0 \forall y_1 \ldots Q y_n \rho(x_0,\ldots,x_{k-1},y_0,\ldots,y_n),$$
where Q is $\exists$ or $\forall$ according as n is even or odd. Similarly, for each Π_{n+1} formula $\pi(x_0,\ldots,x_{k-1})$, there is a PR formula $\rho(x_0,\ldots,x_{k-1},y_0,\ldots,y_n)$ such that
$$PA \vdash \pi(x_0,\ldots,x_{k-1}) \leftrightarrow \forall y_0 \exists y_1 \ldots Q y_n \rho(x_0,\ldots,x_{k-1},y_0,\ldots,y_n),$$
where Q is $\forall$ or $\exists$ according as n is even or odd. (This includes the case k = 0, in which case σ and π are sentences.)

Fact 3(a) can be improved as follows:

Fact 3. (b) The formula $\delta_f(x_0,\ldots,x_n,y)$ of Fact 3(a) can be taken to be Σ_1.

Corollary 1 can now be improved as follows.

Corollary 3. (a) For every recursive set X (relation R) there is a Σ_1 formula and, therefore, a Π_1 formula binumerating X (R) in Q.
(b) For every r. e. set X (relation R) there are a Σ_1 formula and a Π_1 formula numerating X (R) in Q.

A theory T is Γ-*sound* if every Γ sentence provable in T is true.

For every r.e. set X, there is a primitive recursive relation R(k,m) such that X = {k: $\exists$mR(k,m)}. Thus, from Corollary 2(ii), we get the following:

Corollary 4. Suppose T is Σ_1-sound. Then for every r. e. set X, there is a Σ_1 formula numerating X in T.

In Chapter 3 it will be shown that the assumption that T is Σ_1-sound can be omitted (Theorem 3.1).

A function $f(k_0,\ldots,k_n)$ is *provably recursive* in T if there is a Σ_1 formula $\delta_f(x_0,\ldots,x_n,y)$ such that
(i) $T \vdash \delta_f(k_0,\ldots,k_n,y) \leftrightarrow y = f(k_0,\ldots,k_n)$,
(ii) $T \vdash \delta_f(x_0,\ldots,x_n,y) \wedge \delta_f(x_0,\ldots,x_n,z) \rightarrow y = z$,
(iii) $T \vdash \exists y \delta_f(x_0,\ldots,x_n,y)$.
(In (i) $f(k_0,\ldots,k_n)$ is, of course, a numeral.) Thus, all primitive recursive functions are provably recursive in PA.

Suppose (i), (ii), (iii) are true. Then we may add the function symbol f to the language of T and add
$$f(x_0,\ldots,x_n) = y \leftrightarrow \delta_f(x_0,\ldots,x_n,y)$$

as a new axiom, where $\delta_f(x_0,\ldots,x_n,y)$ is as above. The resulting theory is then a conservative extension of T.

Suppose $\alpha(x)$ and $\beta(y)$ are Σ_{n+1} and $f(k_0,\ldots,k_n)$ is provably recursive in T. Then

(1) $\quad \exists x(\alpha(x) \wedge \forall y \leqslant f(x_0,\ldots,x_n)\beta(y))$

is not (necessarily) Σ_{n+1}; it is, however, Σ_{n+1}^T, since it is, provably in T, equivalent to

$\quad \exists xz(\alpha(x) \wedge \delta_f(x_0,\ldots,x_n,z) \wedge \forall y \leqslant z\beta(y))$.

Similarly, if $\alpha(x)$ is Σ_{n+1} and $\beta(y)$ is Π_{n+1}, then

(2) $\quad \forall x(\alpha(x) \to \exists y \leqslant f(x_0,\ldots,x_n)\beta(y))$

is Π_{n+1}^T, since it is, provably in T, equivalent to

$\quad \forall xz(\alpha(x) \wedge \delta_f(x_0,\ldots,x_n,z) \to \exists y \leqslant z\beta(y))$.

(The reason why we don't extend the sets of Σ_{n+1} and Π_{n+1} to comprise the formulas (1) and (2), respectively, is that Σ_{n+1} and Π_{n+1} would then be nonrecursive (see Fact 4(d)).

By Fact 4(b), there is a primitive recursive function $\mathrm{Sbst}_1(k,m,n)$ such that if n is a formula, then $\mathrm{Sbst}_1(k,m,n)$ is the result of replacing v_k in that formula by the numeral for the number m. Thus, if n is $\xi(v_k)$, then $\mathrm{Sbst}_1(k,m,\xi(v_k)) := \xi(m)$. Let

$\quad \mathrm{Sbst}_2(k_0,m_0,k_1,m_1,n) = \mathrm{Sbst}_1(k_0,m_0,\mathrm{Sbst}_1(k_1,m_1,n))$.

By Fact 2, there are formulas $\mathrm{Subst}_1(x,y,z,u)$ and $\mathrm{Subst}_2(x_0,y_0,x_1,y_1,z,u)$ such that

$\quad Q \vdash \mathrm{Subst}_1(k,m,\xi(v_k),u) \leftrightarrow u = \xi(m)$,

$\quad Q \vdash \mathrm{Subst}_2(k_0,m_0,k_1,m_1,\eta(v_{k_0},v_{k_1}),u) \leftrightarrow u = \eta(m_0,m_1)$.

As already mentioned, we may in any extension of PA introduce the corresponding function symbols Sbst_1 and Sbst_2.

If v_k is the only free variable of ξ, we write $\xi(\dot{x})$ for $\mathrm{Sbst}_1(k,x,\xi)$. In writing, for example, $\eta(\dot{x},\dot{y})$ we assume that there are k and m such that $\eta := \eta(v_k,v_m)$ and that $\eta(\dot{x},\dot{y}) := \mathrm{Sbst}_2(k,x,m,y,\eta)$. Note that, although x is not free in, say, $\xi(\eta(x))$, it is free in $\xi(\eta(\dot{x}))$.

Given a formula $\sigma(z)$, let

$\quad \mathrm{Prf}_\sigma(x,y)$

be a formula whose intuitive meaning is: "there is a v such that $(y)_v = x$, $(y)_u = 0$ for all $u > v$, and for every $u \leqslant v$, either $(y)_u$ is a logical axiom, satisfies $\sigma(z)$, or is obtained from formulas $(y)_w$ with $w < u$ using one of the (logical) rules of derivation"; in other words "y is a proof of the sentence x from the set of sentences satisfying $\sigma(z)$". (Thus, if there are nonsentences "satisfying $\sigma(z)$", they are simply disregarded.) The fact that there is a formula $\mathrm{Prf}_\sigma(x,y)$ with the desired properties (see below) follows from Facts 2(i) and 4 (and the details of the formalization of predicate logic.) If $\sigma(z)$ is Γ^+, then $\mathrm{Prf}_\sigma(x,y)$ is Γ^+.

Let
$$\mathrm{Pr}_\sigma(x) := \exists y\,\mathrm{Prf}_\sigma(x,y),$$
$$\mathrm{Con}_\sigma := \neg\mathrm{Pr}_\sigma(\bot).$$

Thus, the intuitive meaning of $\mathrm{Pr}_\sigma(x)$ is: "the sentence x is provable from the set of sentences satisfying $\sigma(z)$" and Con_σ intuitively says: "the set of sentences satisfying $\sigma(z)$ is consistent". If $\sigma(z)$ is Σ_{n+1}, then $\mathrm{Pr}_\sigma(x)$ is Σ_{n+1}, and Con_σ is Π_{n+1}.

For any formula $\sigma(x)$, let
$$(\sigma|y)(x) := \sigma(x) \wedge x \leq y,$$
$$(\sigma + y)(x) := \sigma(x) \vee x = y.$$

In what follows we shall use $\mathrm{Prf}_S(x,y)$, $\mathrm{Prf}_{S+z}(x,y)$, $\mathrm{Prf}_{S|z}(x,y)$, $\mathrm{Pr}_S(x)$, Con_S, etc. to denote (ambiguously) any formula $\mathrm{Prf}_\sigma(x,y)$, $\mathrm{Prf}_{\sigma+z}(x,y)$, $\mathrm{Prf}_{\sigma|z}(x,y)$, $\mathrm{Pr}_\sigma(x)$, Con_σ, etc. where $\sigma(x)$ is a PR binumeration of S. If $S = \emptyset$, we assume that $\sigma(x) := \neg x = x$; if S is finite and nonempty, $S = \{\varphi_0,\ldots,\varphi_n\}$, then $\sigma(x) := x = \varphi_0 \vee \ldots \vee x = \varphi_n$.

Fact 6. $\vdash \forall x(\sigma(x) \to \sigma'(x)) \to \forall y(\mathrm{Pr}_\sigma(y) \to \mathrm{Pr}_{\sigma'}(y))$. Consequently
$\vdash \forall x(\sigma(x) \to \sigma'(x)) \to (\mathrm{Con}_{\sigma'} \to \mathrm{Con}_\sigma)$.

Fact 7. Suppose $\sigma(x)$ numerates S in T.
 (a) If p is a proof of φ in S, then $T \vdash \mathrm{Prf}_\sigma(\varphi,p)$.
 (b) If $S \vdash \varphi$, then $T \vdash \mathrm{Pr}_\sigma(\varphi)$.
 (c) Suppose $PA \dashv T$. Let $\alpha(x_0,\ldots,x_{n-1})$ be any formula whose free variables are $x_0,\ldots,x_{n-1}$. If $S \vdash \alpha(x_0,\ldots,x_{n-1})$, then $T \vdash \mathrm{Pr}_\sigma(\alpha(\dot{x}_0,\ldots,\dot{x}_{n-1}))$.
 (d) If $\sigma(x)$ binumerates S in T and p is not a proof of φ in S, then $T \vdash \neg\mathrm{Prf}_\sigma(\varphi,p)$.

Fact 8. Let $\sigma(x)$ be any formula.
 (i) $PA \vdash \mathrm{Pr}_\sigma(x) \wedge \mathrm{Pr}_\sigma(x \to y) \to \mathrm{Pr}_\sigma(y)$,
 (ii) $PA \vdash \mathrm{Pr}_{\sigma+y}(z) \leftrightarrow \mathrm{Pr}_\sigma(y \to z)$,
 (iii) $PA \vdash \mathrm{Pr}_\sigma(x) \to \exists y\,\mathrm{Pr}_{\sigma|y}(x)$.

Corollary 5. Let $\sigma(x)$ be any formula.
 (i) $PA \vdash \mathrm{Pr}_\sigma(\beta(\dot{x})) \to \mathrm{Pr}_\sigma(\exists x\beta(x))$,
 (ii) $PA \vdash \mathrm{Pr}_\sigma(\forall x\beta(x)) \to \mathrm{Pr}_\sigma(\beta(\dot{x}))$,
 (iii) $PA \vdash \mathrm{Pr}_\sigma(x) \wedge \mathrm{Pr}_\sigma(\neg x) \to \neg\mathrm{Con}_\sigma$,
 (iv) $PA \vdash \mathrm{Pr}_\sigma(\neg x) \leftrightarrow \neg\mathrm{Con}_{\sigma+x}$ and $PA \vdash \mathrm{Pr}_\sigma(x) \leftrightarrow \neg\mathrm{Con}_{\sigma+\neg x}$,

(v) if PA ⊣ T, $\sigma(x)$ numerates S in T, and $S \vdash \gamma(x) \to \delta(x)$, then
$T \vdash Pr_\sigma(\gamma(\dot{x})) \to Pr_\sigma(\delta(\dot{x}))$,
(vi) if PA ⊣ T, $\sigma(x)$ numerates S in T, and $S \vdash \varphi \to \psi$, then
$T \vdash Pr_\sigma(\varphi) \to Pr_\sigma(\psi)$.

All true Σ_1 sentences are provable in Q; in fact, this is provable in PA; in other words, Q is Σ_1-*complete* provably in PA:

Fact 9. Suppose φ and $\delta(x_0,\ldots,x_{n-1})$ are Σ_1.
  (a) If φ true, then $Q \vdash \varphi$.
  (b) $PA \vdash \delta(x_0,\ldots,x_{n-1}) \to Pr_Q(\delta(\dot{x}_0,\ldots,\dot{x}_{n-1}))$;
in particular, $PA \vdash \varphi \to Pr_Q(\varphi)$.
  (c) If $\delta(x_0,\ldots,x_{n-1})$ is PR, there is a primitive recursive function g such that $PA \vdash \delta(x_0,\ldots,x_{n-1}) \to Prf_Q(\delta(\dot{x}_0,\ldots,\dot{x}_{n-1}),f(x_0,\ldots,x_{n-1}))$.

See also Fact 9(d), below.
By Fact 9(a), if ψ is Π_1 and $T + \psi$ is consistent, then ψ is true.

Corollary 6. Suppose $\sigma(x)$ numerates an extension of Q in PA.
  (a) If φ is a Σ_1 sentence, then
  $PA \vdash \varphi \to Pr_\sigma(\varphi)$.
  (b) If $\sigma(x)$ is Σ_1 and $\tau(x)$ is a numeration of T (in PA), then
  $PA \vdash Pr_\sigma(\varphi) \to Pr_\tau(Pr_\sigma(\varphi))$;
in particular, $PA \vdash Pr_\sigma(\varphi) \to Pr_Q(Pr_\sigma(\varphi))$.

The following conditions (cf. Fact 7(b), Fact 8(i), and Corollary 6(b)) are known as the Bernays–Löb provability conditions (for $Pr_T(x)$).

(BLi) if $T \vdash \varphi$, then $PA \vdash Pr_T(\varphi)$,
(BLii) $PA \vdash Pr_T(\varphi) \land Pr_T(\varphi \to \psi) \to Pr_T(\psi)$,
(BLiii) $PA \vdash Pr_T(\varphi) \to Pr_T(Pr_T(\varphi))$.

The construction of "self-referential" sentences and formulas will play a decisive role in what follows. Such constructions are possible in virtue of the following result, the Fixed Point Lemma; we list a number of special cases; a completely general formulation would be needlessly complicated. φ is a *fixed point* of $\xi(x)$ in T if $T \vdash \varphi \leftrightarrow \xi(\varphi)$.

Lemma 1. (a) For any Γ^+ formula $\gamma(x)$, we can effectively find a Γ^+ sentence

φ such that
$$Q \vdash \varphi \leftrightarrow \gamma(\varphi).$$

(b) For any Γ^+ formula $\gamma(x,y)$, we can effectively find a Γ^+ formula $\xi(x)$ such that
$$Q \vdash \xi(x) \leftrightarrow \gamma(x,\xi).$$

(c) For any Γ^+ formulas $\gamma_0(x,y)$ and $\gamma_1(x,y)$, we can effectively find Γ^+ sentences φ_0 and φ_1 such that
$$Q \vdash \varphi_0 \leftrightarrow \gamma_0(\varphi_0,\varphi_1),$$
$$Q \vdash \varphi_1 \leftrightarrow \gamma_1(\varphi_0,\varphi_1).$$

(d) For every Γ^+ formula $\gamma(x,y)$, we can effectively find a Γ^+ formula $\xi(x)$ such that for every k,
$$Q \vdash \xi(k) \leftrightarrow \gamma(k,\xi(k)).$$

(e) Suppose $PA \dashv T$. For every Γ^+ formula $\gamma(x,y)$, we can effectively find a Γ^+ formula $\xi(x)$ such that
$$PA \vdash \xi(x) \leftrightarrow \gamma(x,\xi(\dot{x})).$$

Proof. In what follows x is v_m and y is v_n.

(a) Let
$$\delta(x) := \exists z(\text{Subst}_1(m,x,x,z) \wedge \gamma(z)).$$
We have
$$Q \vdash \text{Subst}_1(m,\delta,\delta,z) \leftrightarrow z = \delta(\delta).$$
It follows that
$$Q \vdash \delta(\delta) \leftrightarrow \gamma(\delta(\delta)).$$
Thus, $\varphi := \delta(\delta)$ is as desired. ◆

(b) Let
$$\eta(x,y) := \exists z(\text{Subst}_1(n,y,y,z) \wedge \gamma(x,z)).$$
We have
$$Q \vdash \text{Subst}_1(n,\eta,\eta,z) \leftrightarrow z = \eta(x,\eta).$$
It follows that
$$Q \vdash \eta(x,\eta) \leftrightarrow \gamma(x,\eta(x,\eta)).$$
Thus, $\xi(x) := \eta(x,\eta)$ is as desired. ◆

(c) For i = 0, 1, let
$$\delta_i(x,y) := \exists z_0 z_1 (\text{Subst}_2(m,x,n,y,x,z_0) \wedge \text{Subst}_2(m,x,n,y,y,z_1) \wedge \gamma_i(z_0,z_1)).$$
We have
$$Q \vdash \text{Subst}_2(m,\delta_0,n,\delta_1,\delta_0,z_0) \leftrightarrow z_0 = \delta_0(\delta_0,\delta_1),$$
$$Q \vdash \text{Subst}_2(m,\delta_0,n,\delta_1,\delta_1,z_1) \leftrightarrow z_1 = \delta_1(\delta_0,\delta_1).$$
It follows that
$$Q \vdash \delta_i(\delta_0,\delta_1) \leftrightarrow \gamma_i(\delta_0(\delta_0,\delta_1),\delta_1(\delta_0,\delta_1)).$$
Thus, $\varphi_0 := \delta_0(\delta_0,\delta_1)$ and $\varphi_1 := \delta_1(\delta_0,\delta_1)$ are as desired. ◆

(d) Let
$$\eta(x,y) := \exists z(\mathrm{Subst}_2(m,x,n,y,y,z) \wedge \gamma(x,z)).$$
We have
$$Q \vdash \mathrm{Subst}_2(m,k,n,\eta,\eta,z) \leftrightarrow z = \eta(k,\eta).$$
It follows that
$$Q \vdash \eta(k,\eta) \leftrightarrow \gamma(k,\eta(k,\eta)).$$
Thus, $\xi(x) := \eta(x,\eta)$ is as desired. ♦

(e) Let
$$\eta(x,y) := \gamma(x,\mathrm{Sbst}_2(m,x,n,y,y)).$$
We have
$$\mathrm{PA} \vdash \mathrm{Sbst}_2(m,x,n,\eta,\eta) = \eta(\dot{x},\eta).$$
It follows that
$$\mathrm{PA} \vdash \eta(x,\eta) \leftrightarrow \gamma(x,\eta(\dot{x},\eta)).$$
Thus, $\xi(x) := \eta(x,\eta)$ is as desired. ∎

The cases listed in the above formulation of the Fixed Point Lemma do not exhaust the possibilities of self-reference. (This should be clear from the proof.) However, applications of self-reference (in what follows) not covered by these examples can be obtained by easy generalization.

For example, let $\gamma(x,y)$ be any formula and suppose we want to construct a sentence θ such that
$$Q \vdash \theta \leftrightarrow \gamma(\theta, \neg\theta).$$
This can be done as follows. There is a (PR) formula $\nu(x,y)$ such that for every φ,
$$Q \vdash \nu(\varphi,y) \leftrightarrow y = \neg\varphi.$$
Let $\delta(x) := \exists y(\nu(x,y) \wedge \gamma(x,y))$. By the Fixed Point Lemma, there is a sentence θ such that $Q \vdash \theta \leftrightarrow \delta(\theta)$. Clearly θ is as desired.

From this point on the Fixed Point Lemma will be used without further mention. The phrase "let φ be such that $Q \vdash \varphi \leftrightarrow \xi(\varphi)$", where $\xi(x)$ is Γ^+, is short for "let φ be a (Γ^+) sentence such that $Q \vdash \varphi \leftrightarrow \xi(\varphi)$" and the same applies *mutatis mutandis* to all similar phrases.

Applying the Fixed Point Lemma we now prove two basic and very important theorems.

A theory S is *decidable* if Th(S) is recursive, otherwise *undecidable*. S is *essentially undecidable* if S and all its consistent extensions are undecidable.

Theorem 2. Q is essentially undecidable.

This follows at once from Corollary 1(a) and:

Lemma 2. There is no formula binumerating Th(T) in T.

Proof. Suppose $\tau(x)$ binumerates Th(T) in T. Let φ be such that
(1) $\quad Q \vdash \varphi \leftrightarrow \neg\tau(\varphi)$.
If $T \vdash \varphi$, then $T \vdash \tau(\varphi)$ and so, by (1), $T \vdash \neg\varphi$. But then T is inconsistent, contrary to Convention 2. It follows that $T \nvdash \varphi$. Since $\tau(x)$ binumerates Th(T) in T, this implies that $T \vdash \neg\tau(\varphi)$ and so, by (1), $T \vdash \varphi$, a contradiction. ∎

Let U be any, not necessarily r.e., consistent extension of Q. A *truth-definition for* U is a formula $\upsilon(x)$ such that for every sentence φ,
(tr) $\quad U \vdash \varphi \leftrightarrow \upsilon(\varphi)$.

The following result is known as the Tarski, or Gödel–Tarski, theorem.

Theorem 3. There is no truth-definition for U.

Proof. The proof is almost the same as that of Lemma 2. Suppose $\upsilon(x)$ is a truth-definition for U. Let φ be such that
$$Q \vdash \varphi \leftrightarrow \neg\upsilon(\varphi).$$
This together with (tr) implies that U is inconsistent, contrary to assumption. ∎

The proof of Theorem 3 is a formal version of the so called Liar paradox. In the latter one considers a sentence saying of itself that it isn't true:
(*) $\quad$ (*) isn't true.
(*) is both true and not true. Thus, a sentence saying, what (*) seems to say, cannot exist.

Thus, there is no full truth-definition in arithmetic. We do, however, have the following partial positive fact. A *partial truth-definition* for Γ sentences in T is a formula $\mathrm{Tr}_\Gamma(x)$ such that for every Γ sentence φ,
$$T \vdash \varphi \leftrightarrow \mathrm{Tr}_\Gamma(\varphi).$$
Let $\Gamma(x)$ be a "natural" PR binumeration of the set of Γ sentences (cf. Fact 4(d)). We (may) assume that
$$PA \vdash \Gamma(x) \to \Gamma^d(\neg x)$$
and that if $\Gamma \subseteq \Gamma'$, then
$$PA \vdash \Gamma(x) \to \Gamma'(x).$$

Fact 10. (a) There is a Γ formula $\mathrm{Sat}_\Gamma(x,y)$ with the following properties:
(i) For every Γ formula $\gamma(x)$,
$$PA \vdash \gamma(x) \leftrightarrow \mathrm{Sat}_\Gamma(x, \gamma).$$
(ii) Let $\mathrm{Tr}_\Gamma(x) := \mathrm{Sat}_\Gamma(0, x)$. Then for every Γ formula $\gamma(x)$,

$$PA \vdash \gamma(x) \leftrightarrow Tr_\Gamma(\gamma(\dot{x}))$$
and so for every Γ sentence φ,
$$PA \vdash \varphi \leftrightarrow Tr_\Gamma(\varphi).$$
(iii) $\quad PA \vdash \Gamma^d(x) \wedge \Gamma(y) \wedge Tr_\Gamma d(x) \wedge Tr_\Gamma(x \to y) \to Tr_\Gamma(y).$

(b) There is a Δ_{n+1} formula $Sat_{B_n}(x,y)$ such that for every B_n formula $\beta(x)$,
$$PA \vdash \beta(x) \leftrightarrow Sat_{B_n}(x,\beta).$$
(iv) Let $Tr_{B_n}(x) := Sat_{B_n}(0,x)$. Then for every B_n formula $\beta(x)$,
$$PA \vdash \beta(x) \leftrightarrow Tr_{B_n}(\beta(\dot{x}))$$
and so for every B_n sentence φ,
$$PA \vdash \varphi \leftrightarrow Tr_{B_n}(\varphi).$$

Fact 10(a)(i), (ii) can be used to justify self-referential constructions such as the following one. Let $\gamma(x,y)$ be any Γ formula. There is then a Γ formula $\xi(x)$ such that
$$PA \vdash \xi(k) \leftrightarrow \xi(k+1) \vee \gamma(k,\xi).$$
Indeed, this is equivalent to
$$PA \vdash \xi(k) \leftrightarrow Tr_\Gamma(\xi(k+1)) \vee \gamma(k,\xi).$$
Fact 9(a), (b) can now be strengthened as follows:

Fact 9. (d) $PA \vdash \Sigma_1(x) \wedge Tr_{\Sigma_1}(x) \to Pr_Q(x).$

Applying Fact 10(a), we can now show that the arithmetical hierarchy, pertaining to formulas of L_A, is proper; for the corresponding result for sentences, see Corollary 2.6.

Theorem 4. Suppose $PA \dashv T$.
(a) There is a Γ formula which is not $\Gamma^{d,T}$.
(b) There is a Δ_{n+1} formula which is not B_n^T.

Proof. (a) Let $\gamma(x) := Sat_\Gamma(x,x)$. Suppose $\eta(x)$ is Γ^d and $T \vdash \eta(x) \leftrightarrow \gamma(x)$. $\neg\eta(x)$ is Γ. Thus, $T \vdash \neg\eta(x) \leftrightarrow Sat_\Gamma(x,\neg\eta)$ and so $T \vdash \neg\eta(\neg\eta) \leftrightarrow \gamma(\neg\eta)$. But also $T \vdash \eta(\neg\eta) \leftrightarrow \gamma(\neg\eta)$. It follows that $T \vdash \neg\eta(\neg\eta) \leftrightarrow \eta(\neg\eta)$. But then T is inconsistent, contrary to Convention 2. ♦
(b) Similar. ∎

In terms of the partial truth-definitions we can formulate the following:

Fact 11. For every Γ,
$$PA \vdash \forall x(\Gamma(x) \wedge Pr_\emptyset(x) \to Tr_\Gamma(x)).$$

Let $X|k = \{n \in X : n \leq k\}$. A theory T is *reflexive* if $T \vdash \text{Con}_{T|k}$ for every k. T is *essentially reflexive* if every extension of T (in the same language) is reflexive.

Corollary 7. PA is essentially reflexive.

Proof. Suppose PA ⊣ T. Let k be arbitrary and let Γ be such that $\neg \wedge T|k$ is Γ. By Fact 10(a)(ii) and Fact 11, $T \vdash \text{Pr}_\emptyset(\neg \wedge T|k) \to \neg \wedge T|k$ and so $T \vdash \neg \text{Pr}_\emptyset(\neg \wedge T|k)$. But then, by Corollary 5(iv), $T \vdash \text{Con}_{T|k}$, as desired. ∎

From Facts 6 and 11 and Corollaries 5(iv) and 7 we get:

Corollary 8. Suppose PA ⊣ T.
 (a) If $\tau(x)$ binumerates T in T, then $T \vdash \text{Con}_{\tau|k}$ for every k.
 (b) For all k and φ, $T \vdash \text{Pr}_{T|k}(\varphi) \to \varphi$.

If we assume, as we may, that for every Γ formula $\xi(x)$, $PA \vdash \Gamma(\xi(\dot{x}))$, then from Fact 11 we also get the following:

Corollary 8. (c) For every finite subtheory S of T and every formula $\xi(x)$,
$$T \vdash \text{Pr}_S(\xi(\dot{x})) \to \xi(x).$$

Corollary 7 will be of crucial importance, especially in Chapters 6 and 7. But it should be observed that, although many results proved in the following pages for extensions of PA do depend on Corollary 7, others, for example, most of those of Chapter 2 and all results of Chapter 5, do not. The latter results generalize to (possibly finitely) axiomatized, consistent extensions of PA, not (necessarily) formalized in L_A.

The following elementary observations are occasionally useful.

Lemma 3. Let
$$\pi := \forall x(\alpha(x) \to \exists y \leq x \beta(y)), \quad \theta := \forall y(\beta(y) \to \exists x < y \alpha(x)),$$
$$\sigma := \exists x(\alpha(x) \wedge \forall y \leq x \neg \beta(y)), \quad \chi := \exists y(\beta(y) \wedge \forall x < y \neg \alpha(x)).$$
Then
 (i) $PA \vdash \pi \vee \theta$,
 (ii) $PA \vdash \neg(\sigma \wedge \chi)$,
 (iii) $PA \vdash (\pi \wedge \theta) \to \forall x \neg \alpha(x)$,
 (iv) $PA \vdash \exists x \alpha(x) \to (\sigma \vee \chi)$,
 (v) $PA \vdash \sigma \leftrightarrow (\exists x \alpha(x) \wedge \theta)$,
 (vi) $PA \vdash \exists x \alpha(x) \to (\chi \leftrightarrow \neg \sigma)$.

Proof. (i) Argue in PA: "Suppose $\neg\pi$ and $\neg\theta$. Let z and u be such that $\alpha(z)$, $\neg\exists y{\leqslant}z\,\beta(y)$, $\beta(u)$, $\neg\exists y{<}u\,\alpha(y)$. Then $\neg u \leqslant z$ and $\neg z < u$, impossible. Thus, $\pi \vee \theta$." This proves (i).

(ii) follows from (i).

(iii) Argue in PA: "Suppose π, θ, and $\exists x\,\alpha(x)$. By the least number principle, there is a smallest z such that $\alpha(z)$. Since π holds, there is a $u \leqslant z$ such that $\beta(u)$. But then, by θ, there is a $v < u$ such that $\alpha(v)$. It follows that $v < z$, a contradiction." This proves (iii).

(iv) follows from (iii).

(v) follows from (i) and (iii).

(vi) follows from (ii) and (iv). ∎

Corollary 9. Suppose $\alpha(x)$ and $\beta(y)$ are PR. Let θ be as in Lemma 3.

(a) If $\exists x\,\alpha(x)$ and $\forall y\,\neg\beta(y)$ are true, then $PA \vdash \theta$.

(b) $PA \vdash (\exists x\,\alpha(x) \wedge \forall y\,\neg\beta(y)) \to Pr_{PA}(\theta)$.

Proof. Let π be as in Lemma 3.

(a) If $\exists x\,\alpha(x)$ and $\forall x\,\neg\beta(x)$ are true, so is $\neg\pi$. Since $\neg\pi$ is Σ_1, it follows that $PA \vdash \neg\pi$. But then, by Lemma 3(i), $PA \vdash \theta$. ♦

(b) $\vdash (\exists x\,\alpha(x) \wedge \forall y\,\neg\beta(y)) \to \neg\pi$. Since $\neg\pi$ is Σ_1, we have $PA \vdash \neg\pi \to Pr_{PA}(\neg\pi)$, by Fact 9(b). By Lemma 3(i), $PA \vdash \neg\pi \to \theta$. By (BLi) and (BLii), we get $PA \vdash Pr_{PA}(\neg\pi) \to Pr_{PA}(\theta)$. Putting these together we get the desired conclusion. ∎

For the concepts and results of (elementary) recursion theory used in this book we refer to Soare (1987). A set is Π_n^0 (Σ_n^0) if it is defined in **N** by a Π_n (Σ_n) formula. X is a *complete* Π_n^0 (Σ_n^0) set if X is Π_n^0 (Σ_n^0) and for every Π_n^0 (Σ_n^0) set Y, there is a recursive function f(k) such that for every k, $k \in Y$ iff $f(k) \in X$. Complete Π_n^0 (Σ_n^0) sets exist. No complete Π_n^0 (Σ_n^0) set is Σ_n^0 (Π_n^0). If T is true, Theorem 4 follows directly from the fact that for each $n > 0$, there is a Π_n^0 (Σ_n^0) set which isn't Σ_n^0 (Π_n^0).

The following notions will be needed in Chapter 7. A partially ordered set $\mathbf{L} = (L, \leqslant)$ is a *lattice* if any two members a, b of L have a *least upper bound* (l.u.b.) $a \cup b$ and a *greatest lower bound* (g.l.b.) $a \cap b$. Thus, $a \cup b \leqslant c$ iff $a \leqslant c$ and $b \leqslant c$; similarly, $c \leqslant a \cap b$ iff $c \leqslant a$ and $c \leqslant b$. It follows that $a \leqslant b$ iff $a \cap b = a$ iff $a \cup b = b$. **L** is *distributive* if for all a, b, $c \in L$,

$$a \cap (b \cup c) = (a \cap b) \cup (a \cap c);$$

or equivalently,

$$a \cup (b \cap c) = (a \cup b) \cap (a \cup c).$$

The inequalities

$$(a \cap b) \cup (a \cap c) \leqslant a \cap (b \cup c),$$
$$a \cup (b \cap c) \leqslant (a \cup b) \cap (a \cup c)$$
hold in all lattices.

Suppose **L** has a minimal (maximal) element 0_L (1_L). If $a \cap b = 0_L$ and $a \cup b = 1_L$, then b is a *complement* of a (and a a complement of b). If **L** is distributive, each a has at most one complement. If $b = \max\{c: a \cap c = 0_L\}$, then b is the *pseudocomplement (p.c.)* of a.

Notes for Chapter 1

The background material on formal arithmetic and Gödel numberings presupposed in this book can be found in several textbooks, for example, Kleene (1952a), Mendelson (1987), Smoryński (1985), Kaye (1991), Boolos (1979), (1993), Hájek and Pudlák (1993). We follow Feferman (1960) in identifying formal expressions with their Gödel numbers. Facts 2 and 4 are due to Gödel (1931). When, somewhat later, the recursive functions were defined, the proof of Fact 3 presented no new difficulties. For proofs of Facts 1–4, see any one of the textbooks just mentioned. The terms "numerate" and "binumerate" are due to Feferman (1960). Theorem 1 is due to Craig (1953). The exact definitions of PR, Σ_n, Π_n, Δ_n vary from one author to another, depending on the intended applications; for example, other authors often use Δ_0 to denote the set of bounded formulas, i.e., formulas all of whose quantifiers are bounded (cf. e.g. Kaye (1991) and Hájek and Pudlák (1993)). The present definitions have the advantage that the concepts are easy to work with (in the present setting) and that the sets PR, Σ_n, Π_n are primitive recursive (Fact 4(d)) (Δ_n, $n > 0$, is not recursive; see Exercise 2.4). The formulas $\mathrm{Prf}_\sigma(x,y)$, $\mathrm{Pr}_\sigma(x)$, Con_σ were introduced by Feferman (1960); that the meaning and properties in terms of provability of, for example, Con_σ depends very much on the exact choice of $\sigma(x)$, and not just on the set (bi)numerated by $\sigma(x)$, was stressed by Feferman (see e.g. Theorems 2.7 and 2.8). Fact 9 is due to Feferman (1960). The Bernays–Löb provability conditions are due to Löb (1955), simplifying the original conditions due to Bernays (cf. Hilbert and Bernays (1939)). Part (a) of the Fixed Point Lemma is implicit in Gödel (1931); it was first stated explicitly by Carnap (1934) (see also Gödel (1934)). The more general versions (b)–(e) were subsequently obtained by Ehrenfeucht and Feferman (1960) and Montague (1962). Lemma 2 and Theorem 2 first appeared in Tarski, Mostowski, Robinson (1953). Theorem 3 was first published by

Tarski (1933) (see also Gödel (1934)). The application of (partial) truth-definitions goes back to Hilbert and Bernays (1934, 1939); a full proof of Fact 10 is given in Kaye (1991). Fact 11 is essentially due to Kreisel and Wang (1955) (see also Mostowski (1952a)); for a sketch of a proof of a related result, which can easily be turned into a proof of Fact 11, see Kaye (1991), p. 140. Corollary 7 is due to Mostowski (1952a).

2. INCOMPLETENESS

The methods of arithmetization and self-reference were originally used to prove incompleteness theorems for arithmetical theories. In this chapter we present the most important theorems of this type.

A sentence φ (in the language of S) is *undecidable* in S if $S \nvdash \varphi$ and $S \nvdash \neg\varphi$. S is *complete* if no sentence is undecidable in S, otherwise *incomplete*.

§1. Incompleteness. We begin with the first and most important result of the whole subject, Gödel's incompleteness theorem (for theories in L_A).

Theorem 1. Let φ be a Π_1 sentence such that
(G) $\quad Q \vdash \varphi \leftrightarrow \neg\Pr_T(\varphi)$.
Then φ is true and $T \nvdash \varphi$. Thus, if T is Σ_1-sound, then also $T \nvdash \neg\varphi$.

Proof. Suppose $T \vdash \varphi$. Then, by Fact 7(b), $Q \vdash \Pr_T(\varphi)$. But then, by (G), $Q \vdash \neg\varphi$ and so $T \vdash \neg\varphi$. It follows that T is inconsistent, contrary to Convention 2. Thus, $T \nvdash \varphi$. By (G), φ is true. Thus, $\neg\varphi$ is a false Σ_1 sentence and so $T \nvdash \neg\varphi$ if T is Σ_1-sound. ∎

Notice the close similarity between the proofs of Theorem 1, Lemma 1.2, and Theorem 1.3 (the Liar paradox).

To derive the conclusion that $T \nvdash \neg\varphi$ in Theorem 1, we needed the assumption that T is Σ_1-sound. We can now see that this is stronger than mere consistency: $T + \neg\varphi$ is consistent but not Σ_1-sound. (Note that it does not follow from Theorem 1 that $T + \neg\varphi$ is incomplete.) Thus, the question arises if, assuming consistency only, there is a (Π_1) sentence which is undecidable in T. Our next result, known as Rosser's theorem, shows that the answer is affirmative.

Theorem 2. Let θ be a Π_1 sentence such that
(R) $\quad Q \vdash \theta \leftrightarrow \forall z(\Prf_T(\theta,z) \rightarrow \exists u \leqslant z \Prf_T(\neg\theta,u))$.
Then θ is undecidable in T.

Proof. We first prove that $T \nvdash \theta$. Suppose, for *reductio ad absurdum*, $T \vdash \theta$ and let p be a proof of θ in T. Then, by Fact 7(a),
(1) $\quad Q \vdash \Prf_T(\theta,p)$.

§1. *Incompleteness*

Since T is consistent, we have $T \nvdash \neg\theta$. By Fact 7(d), $Q \vdash \neg\text{Prf}_T(\neg\theta,q)$ for every q. But then, by Fact 1(iv),
$$Q \vdash u \leq p \to \neg\text{Prf}_T(\neg\theta,u).$$
Combining this with (1) we get
$$Q \vdash \exists z(\text{Prf}_T(\theta,z) \wedge \forall u \leq z \neg\text{Prf}_T(\neg\theta,u)).$$
But then, by (R), $Q \vdash \neg\theta$ and so $T \vdash \neg\theta$, a contradiction. Thus, $T \nvdash \theta$ as desired.

Next we prove that $T \nvdash \neg\theta$. Suppose $T \vdash \neg\theta$ and let p be a proof of $\neg\theta$ in T. Then $T \nvdash \theta$ and so, by Fact 7(d), $Q \vdash \neg\text{Prf}_T(\theta,q)$ for every q. By Fact 1(iv), it follows that
$$Q \vdash z < p \to \neg\text{Prf}_T(\theta,z),$$
whence, by Fact 1(v),
(2) $\qquad Q \vdash \text{Prf}_T(\theta,z) \to p \leq z.$
By Fact 7(a), $Q \vdash \text{Prf}_T(\neg\theta,p)$. Hence, trivially,
$$Q \vdash p \leq z \to \exists u \leq z \text{Prf}_T(\neg\theta,u).$$
Combining this with (2) and (R) we get $Q \vdash \theta$ and so $T \vdash \theta$, again a contradiction. It follows that $T \nvdash \neg\theta$, as desired. ∎

Arguments similar to the above proof will occur time and again in the following pages.

Theorem 2 can also be proved by considering a Σ_1 sentence ψ such that
(R') $\qquad Q \vdash \psi \leftrightarrow \exists z(\text{Prf}_T(\neg\psi,z) \wedge \forall u \leq z \neg\text{Prf}_T(\psi,u)),$
a condition that is, of course, (almost) satisfied by $\neg\theta$, where θ is as in (R). A sentence satisfying (R) or (R') is called a *Rosser sentence* for T.

The difference between (the proofs of) Theorems 1 and 2 can be described in the following way. The formula $\xi(x) := \text{Pr}_T(x)$ used in the former has the properties: (i) if $T \vdash \varphi$, then $T \vdash \xi(\varphi)$, and (ii) if $T \vdash \neg\varphi$, then ($T \nvdash \varphi$ and so) $\xi(\varphi)$ is false. The corresponding formula which is (almost) used in the latter,
$$\xi(x) := \exists z(\text{Prf}_T(x,z) \wedge \forall u \leq z \neg\text{Prf}_T(\neg x,u)),$$
satisfies (i) and (iii): if $T \vdash \neg\varphi$, then $T \vdash \neg\xi(\varphi)$. From (i) and (iii) it follows at once that if
$$T \vdash \psi \leftrightarrow \neg\xi(\psi)$$
(or $T \vdash \psi \leftrightarrow \xi(\neg\psi)$), then ψ is undecidable in T.

If $PA \dashv T$, the above proof of Theorem 2 can be replaced by the following argument. Suppose $T \vdash \theta$. Then $T \nvdash \neg\theta$. By (R), it follows that $\neg\theta$ is true and so, $\neg\theta$ being Σ_1, $T \vdash \neg\theta$, by Fact 9(a), a contradiction. (This part does not use the assumption that $PA \dashv T$.) Next suppose $T \vdash \neg\theta$. Then $T \nvdash \theta$. But then, by Corollary 1.9(a) and (R), $T \vdash \theta$, again a contradiction.

That T is incomplete also follows from Theorem 1.2, since every complete r.e. theory is decidable. This proof, however, does not (directly) yield an example of a sentence undecidable in T. Furthermore, the present proof of Theorem 1 is needed in the proof of Theorem 4, below.

That every complete r.e. theory U is decidable is seen as follows: If U is inconsistent, decidability is trivial; thus, suppose U is consistent. Let φ be any sentence of U. To decide whether or not $\varphi \in \text{Th}(U)$, generate, in some effective way, all proofs in U. If a proof of φ is found, conclude that $\varphi \in \text{Th}(U)$; if a proof of $\neg\varphi$ is found, conclude that $\varphi \notin \text{Th}(U)$.

Conversely, Theorem 1.2 follows from Theorem 2. Indeed, suppose U is a consistent, decidable extension of Q. There is then a complete, recursive, consistent extension U' of U. U' is an extension of Q. Hence, by Craig's theorem (Theorem 1.1), there is a complete, consistent primitive recursive extension of Q. This, however, contradicts Theorem 2.

That any consistent, decidable theory U has a complete, consistent, decidable extension can be seen as follows: Let $\varphi_0, \varphi_1, \ldots$ be an effective enumeration of all sentences of the language of U. Define U_n by: $U_0 = U$, $U_{n+1} = U_n + \varphi_n$ if $U_n \nvdash \neg\varphi_n$, $U_{n+1} = U_n + \neg\varphi_n$ otherwise. Let $U' = \bigcup \{U_n : n \in N\}$. Then U' is complete and consistent. By assumption, it can be effectively decided whether $U_n \vdash \neg\varphi_n$ or not. It follows that U' is decidable.

Theorem 2 can be strengthened as follows. A family $\{T_k : k \in N\}$ of theories is *r.e.* if the binary relation $\varphi \in T_k$ is r.e.

Theorem 3. If $\{T_k : k \in N\}$ is an r.e. family of extensions of Q, there is a Π_1 sentence which is simultaneously undecidable in all the theories T_k.

We derive this from the following slight improvement of Theorem 2.

Let us say that a set X of sentences is *monoconsistent with* T if $T + \varphi$ is consistent for every $\varphi \in X$. Thus, for example, if φ is undecidable in T, then $\{\varphi, \neg\varphi\}$ is monoconsistent with T. Also, if X and Y are monoconsistent with T, so is $X \cup Y$. Let $\varphi^0 := \varphi$ and $\varphi^1 := \neg\varphi$.

Lemma 1. If X is r.e. and monoconsistent with Q, there is a Π_1 sentence θ such that $\theta^i \notin X$, $i = 0, 1$.

Proof. The proof is almost the same as the proof of Rosser's theorem. Let $R(k,m)$ be a primitive recursive relation such that $X = \{k: \exists m R(k,m)\}$ and let $\rho(x,y)$ be a PR binumeration of $R(k,m)$. Let θ be such that
(1) $Q \vdash \theta \leftrightarrow \forall z(\rho(\theta,z) \rightarrow \exists u \leqslant z \rho(\neg\theta,u))$.
Suppose either $\theta \in X$ or $\neg\theta \in X$. Let m be the smallest number such that

either $R(\theta,m)$ or $R(\neg\theta,m)$. Suppose first $R(\neg\theta,m)$. Then $\neg\theta \in X$. Also not $R(\theta,n)$ and so $Q \vdash \neg\rho(\theta,n)$ for $n < m$. It follows, by Fact 1(v), that $Q \vdash \rho(\theta,z) \to m \leq z$. Now $Q \vdash \rho(\neg\theta,m)$ and so

$$Q \vdash \forall z(\rho(\theta,z) \to \exists u \leq z \rho(\neg\theta,u)).$$

But then, by (1), $Q \vdash \theta$ which is impossible, since $\neg\theta \in X$.

Thus, not $R(\neg\theta,m)$ and so $R(\theta,m)$ whence $\theta \in X$. Also not $R(\neg\theta,n)$ for $n \leq m$. It follows that $Q \vdash \rho(\theta,m)$ and, by Fact 1(iv), $Q \vdash u \leq m \to \neg\rho(\neg\theta,u)$. But then

$$Q \vdash \exists z(\rho(\theta,z) \land \forall u \leq z \neg\rho(\neg\theta,u))$$

and so, by (1), $Q \vdash \neg\theta$, which is impossible, since $\theta \in X$. Thus, we have derived the desired contradiction and the proof is complete. ∎

Proof of Theorem 3. The set $\bigcup\{\text{Th}(T_k): k \in \mathbb{N}\}$ is r.e. and monoconsistent with Q. Now use Lemma 1. ∎

§2. Consistency statements.

Most arguments carried out in this book can be formalized in PA. In particular this is true of the proof of Theorem 1. This leads to a proof of the following very important result, Gödel's second incompleteness theorem (for theories in L_A). (Recall that a *numeration* of a set X is a formula numerating X in PA.)

Theorem 4. Suppose PA ⊣ T.
 (a) Let φ be as in (G). Then $PA \vdash \text{Con}_T \to \varphi$ and consequently $T \nvdash \text{Con}_T$.
 (b) If $\tau(x)$ is any Σ_1 formula numerating (an extension of) T in T, then $T \nvdash \text{Con}_\tau$.

Proof. (a) We follow closely the proof of Theorem 1(a). By (BLiii),
(1) $PA \vdash \text{Pr}_T(\varphi) \to \text{Pr}_T(\text{Pr}_T(\varphi))$.
By (G) and (BLi), $PA \vdash \text{Pr}_T(\text{Pr}_T(\varphi) \to \neg\varphi)$ and so, by (BLii),

$$PA \vdash \text{Pr}_T(\text{Pr}_T(\varphi)) \to \text{Pr}_T(\neg\varphi).$$

But then, by (1),

$$PA \vdash \text{Pr}_T(\varphi) \to \text{Pr}_T(\neg\varphi),$$

whence, by Corollary 1.5(iii), $PA \vdash \text{Pr}_T(\varphi) \to \neg\text{Con}_T$ and so, by (G),

$$PA \vdash \text{Con}_T \to \varphi.$$

But then, assuming that $T \vdash \text{Con}_T$, we get $T \vdash \varphi$, contradicting Theorem 1(a). It follows that $T \nvdash \text{Con}_T$. ♦
 (b) Similar. ∎

In applying Theorem 4 to an extension S of PA (not assumed to be consistent), we often show that there is a PR binumeration (Σ_1 numeration)

$\sigma(x)$ of S such that $S \vdash \text{Con}_\sigma$ and conclude that S is inconsistent.

A somewhat shorter proof of Theorem 4(a) is as follows. By (G),
$$PA \vdash \neg\varphi \to \Pr_T(\varphi).$$
By provable Σ_1-completeness (Fact 9(b)),
$$PA \vdash \neg\varphi \to \Pr_T(\neg\varphi).$$
But then, by Corollary 1.5(iii), $PA \vdash \neg\varphi \to \neg\text{Con}_T$ and so
$$PA \vdash \text{Con}_T \to \varphi.$$
A similar proof yields Theorem 4(b).

Combining Theorem 4 and Corollary 1.7, we get:

Corollary 1. If $PA \dashv T$, then T is not finitely axiomatizable.

Proof. Suppose T is finitely axiomatizable. Then there is a k such that $T \dashv T|k$. Also, by Corollary 1.7, $T \vdash \text{Con}_{T|k}$, whence $T|k \vdash \text{Con}_{T|k}$. But, since $PA \dashv T|k$, this contradicts Theorem 4. ∎

Corollary 1 will be strengthened in Chapter 4 (Corollary 4.1) and Chapter 6 (Theorem 6.3).

The proof of Theorem 4 can also be formalized in PA yielding:

Corollary 2. If $PA \dashv T$, then $PA + \text{Con}_T \vdash \text{Con}_{T+\neg\text{Con}_T}$.

Proof. Let φ be as in (G). By Theorem 4(a),
(1) $PA \vdash \text{Con}_T \to \varphi.$
But then, by (BLi) and (BLii), $PA \vdash \Pr_T(\text{Con}_T) \to \Pr_T(\varphi)$ and so, by (G)
(2) $PA \vdash \Pr_T(\text{Con}_T) \to \neg\varphi.$
From (1) and (2) we get $PA \vdash \Pr_T(\text{Con}_T) \to \neg\text{Con}_T$ which, by Corollary 1.5(iv), yields the desired conclusion. ∎

The proof of our next result is another exercise in formalization, in this case of the proof of Theorem 2.

Theorem 5. Let θ be a Rosser sentence for T. Then
$$PA + \text{Con}_T \vdash \neg\Pr_T(\theta) \wedge \neg\Pr_T(\neg\theta).$$

Proof. We follow closely the above proof of Theorem 2. By Corollary 1.5(iii),
(1) $PA + \text{Con}_T \vdash \Pr_T(\theta) \to \neg\Pr_T(\neg\theta).$
It follows that
$$PA + \text{Con}_T \vdash \text{Prf}_T(\theta,z) \to \neg\text{Prf}_T(\neg\theta,u)$$
and so

§2. Consistency Statements

(2) $\quad \text{PA} + \text{Con}_T \vdash \text{Prf}_T(\theta, z) \to \forall u \leq z \neg \text{Prf}_T(\neg\theta, u))$.

Let
$$\gamma(z) := \text{Prf}_T(\theta, z) \wedge \forall u \leq z \neg \text{Prf}_T(\neg\theta, u)).$$
Then, by (2),

(3) $\quad \text{PA} + \text{Con}_T \vdash \text{Prf}_T(\theta, z) \to \gamma(z)$.

By Fact 9(b), we have, $\text{PA} \vdash \gamma(z) \to \text{Pr}_T(\gamma(\dot{z}))$. Combining this with (3) yields
$$\text{PA} + \text{Con}_T \vdash \text{Prf}_T(\theta, z) \to \text{Pr}_T(\gamma(\dot{z}))$$
whence, by Corollary 1.5(i),

(4) $\quad \text{PA} + \text{Con}_T \vdash \text{Pr}_T(\theta) \to \text{Pr}_T(\exists z \gamma(z))$.

By (R), $T \vdash \exists z \gamma(z) \to \neg\theta$. But then, by (BLi) and (BLii),
$$\text{PA} \vdash \text{Pr}_T(\exists z \gamma(z)) \to \text{Pr}_T(\neg\theta).$$
Combining this with (4), we get $\text{PA} + \text{Con}_T \vdash \text{Pr}_T(\theta) \to \text{Pr}_T(\neg\theta)$. But then, by (1),

(5) $\quad \text{PA} + \text{Con}_T \vdash \neg\text{Pr}_T(\theta)$,

as desired.

Next we prove that

(6) $\quad \text{PA} + \text{Con}_T \vdash \neg\text{Pr}_T(\neg\theta)$.

From (1), we get
$$\text{PA} + \text{Con}_T \vdash \text{Prf}_T(\neg\theta, u) \to \neg\text{Prf}_T(\theta, z)$$
and so
$$\text{PA} + \text{Con}_T \vdash \text{Prf}_T(\neg\theta, u) \to \forall z < u \neg \text{Prf}_T(\theta, z).$$
Let
$$\delta(u) := \text{Prf}_T(\neg\theta, u) \wedge \forall z < u \neg \text{Prf}_T(\theta, z).$$
By an argument similar to the proof of (4), we get
$$\text{PA} + \text{Con}_T \vdash \text{Pr}_T(\neg\theta) \to \text{Pr}_T(\exists u \delta(u)).$$
(R) easily implies that $T \vdash \exists u \delta(u) \to \theta$. But then (6) follows, by an argument almost the same as the proof of (5). ∎

If $\text{PA} \dashv T$, this proof of Theorem 5 can be replaced by the formalization of the above short proof of Theorem 2. By (R),
$$\text{PA} \vdash \text{Pr}_T(\theta) \wedge \neg\text{Pr}_T(\neg\theta) \to \neg\theta.$$
Since $\neg\theta$ is Σ_1, $\text{PA} \vdash \neg\theta \to \text{Pr}_T(\neg\theta)$. It follows that $\text{PA} \vdash \text{Pr}_T(\theta) \to \text{Pr}_T(\neg\theta)$ and so, by Corollary 1.5(iii),
$$\text{PA} \vdash \text{Con}_T \to \neg\text{Pr}_T(\theta).$$
Next, by Corollary 1.9(b), (R), (BLi), and (BLii), $\text{PA} \vdash \text{Pr}_T(\neg\theta) \wedge \neg\text{Pr}_T(\theta) \to \text{Pr}_T(\theta)$, whence $\text{PA} \vdash \text{Pr}_T(\neg\theta) \to \text{Pr}_T(\theta)$ and so, by Corollary 1.5(iii),
$$\text{PA} \vdash \text{Con}_T \to \neg\text{Pr}_T(\neg\theta).$$
Combining Theorem 5 and Corollary 1.5(iv) we get:

Corollary 3. Let θ be as in (R). Then
$$PA + Con_T \vdash Con_{T+\theta} \wedge Con_{T+\neg\theta}.$$

The sentence φ in (G) above says of itself that it is not provable in T. Let us now consider a sentence χ saying of itself that it *is* provable in T, i.e., such that
$$Q \vdash \chi \leftrightarrow Pr_T(\chi).$$
Is χ provable in T? In this case no simple argument in terms truth will yield an answer, not even if T is true. Nevertheless, it turns out that $T \vdash \chi$ provided that PA ⊣ T. This follows from our next result, known as Löb's theorem.

Theorem 6. Suppose PA ⊣ T and let φ be any sentence such that $T \vdash Pr_T(\varphi) \to \varphi$. Then $T \vdash \varphi$.

Proof. Let θ be such that
(L) $PA \vdash \theta \leftrightarrow (Pr_T(\theta) \to \varphi)$.
From this, (BLi), and (BLii), we get
(1) $PA \vdash Pr_T(\theta) \to (Pr_T(Pr_T(\theta)) \to Pr_T(\varphi))$.
By (BLiii),
(2) $PA \vdash Pr_T(\theta) \to Pr_T(Pr_T(\theta))$.
From (1) and (2) it follows that
(3) $PA \vdash Pr_T(\theta) \to Pr_T(\varphi)$.
Since, by hypothesis, $T \vdash Pr_T(\varphi) \to \varphi$, this implies that
(4) $T \vdash Pr_T(\theta) \to \varphi$.
But then, by (L), $T \vdash \theta$, whence, by (BLi), $PA \vdash Pr_T(\theta)$. Finally, this together with (4) yields $T \vdash \varphi$, as desired. ∎

There is a semantic paradox related to the above proof in somewhat the same way as the Liar paradox is related to the proof of Theorem 1. Let
(**)   If (**) is true, the earth is flat.
"Prove", by considering (**), that the earth is flat.

Theorem 6 is a strengthening of Theorem 4: let $\varphi := \bot$. But Theorem 6 can also be derived from Theorem 4 as follows. Suppose $T \vdash Pr_T(\varphi) \to \varphi$. Then $T + \neg\varphi \vdash \neg Pr_T(\varphi)$, whence, by Corollary 1.5(iv), $T + \neg\varphi \vdash Con_{T+\neg\varphi}$. But then, by Theorem 4, $T + \neg\varphi$ is inconsistent and so $T \vdash \varphi$.

By slightly modifying the proof of Theorem 6 we can derive the stronger result that for every sentence φ,
(L2) $PA \vdash Pr_T(Pr_T(\varphi) \to \varphi) \to Pr_T(\varphi)$.
In fact, from (3) we get
$$PA \vdash (Pr_T(\varphi) \to \varphi) \to (Pr_T(\theta) \to \varphi).$$

§2. *Consistency Statements*   33

But then, by (L), $PA \vdash (Pr_T(\varphi) \to \varphi) \to \theta$, whence, by (BLi) and (BLii),
$$PA \vdash Pr_T(Pr_T(\varphi) \to \varphi) \to Pr_T(\theta).$$
Finally, (L2) follows from this and (3).

Theorem 4 is sometimes informally expressed by saying that if T is as assumed, then T does not prove that T is consistent. That this must be interpreted with some care is clear from the following result.

Theorem 7. Suppose $PA \dashv T$. Let $\tau(x)$ be any formula binumerating T in T and let
$$\tau^*(x) := \tau(x) \wedge Con_{\tau|x}.$$
Then (i) $\tau^*(x)$ binumerates T in T and (ii) $PA \vdash Con_{\tau^*}$.

The following intuitive proof of Theorem 7(ii) (formalizable in PA) is probably easier to understand than the formal argument below, but its formalization would be somewhat longer: "Any proof p from the set X defined by $\tau(x) \wedge Con_{\tau|x}$ contains a greatest sentence $\varphi \in X$. Since φ satisfies $Con_{\tau|x}$, it follows that the set of members of X occurring in p is consistent. Thus, p cannot be a proof of $\bot$."

Proof of Theorem 7. Note that x is free in $Con_{\tau|x}$.
  (i) If $k \in T$, then $T \vdash \tau(k)$. By Corollary 1.8(a), $T \vdash Con_{\tau|k}$. Thus, $T \vdash \tau^*(k)$. If, on the other hand, $k \notin T$, then $T \vdash \neg\tau(k)$ and so $T \vdash \neg\tau^*(k)$.
  (ii) Trivially $\vdash \tau^*(x) \to \tau(x)$. Hence, by Fact 6,
(1) $\vdash Con_\tau \to Con_{\tau^*}$.
Since PA is reflexive, we have $PA \vdash Con_{\tau|0}$. (We assume that 0 is not a formula.) Also, by Fact 8(iii), $PA \vdash \forall z Con_{\tau|z} \to Con_\tau$. By the least number principle, it follows that
(2) $PA \vdash \neg Con_\tau \to \exists z(\neg Con_{\tau|z+1} \wedge Con_{\tau|z})$.
By Fact 6,
$$PA \vdash \neg Con_{\tau|z+1} \to (Con_{\tau|x} \to x \leq z).$$
Hence, by the definition of $\tau^*(x)$,
$$PA \vdash \neg Con_{\tau|z+1} \to (\tau^*(x) \to \tau(x) \wedge x \leq z).$$
Hence, again by Fact 6,
$$PA \vdash \neg Con_{\tau|z+1} \wedge Con_{\tau|z} \to Con_{\tau^*}.$$
But then, by (2),
$$PA \vdash \neg Con_\tau \to Con_{\tau^*}$$
and so, by (1), $PA \vdash Con_{\tau^*}$, as desired. ∎

If $\tau(x)$ is PR, then $\tau^*(x)$ is Π_1. By Theorems 4 and 7, $\tau^*(x)$ is not provably in T equivalent to a Σ_1 formula.

The formula $\tau^*(x)$ may seem like a mere curiosity, but certain closely related formulas are actually of crucial importance in connection with interpretability (see the proof of Lemma 6.2.).

By Theorem 7, the assumption in Theorem 4(b) that $\tau(x)$ is Σ_1 cannot be omitted. But we do have the following:

Corollary 4. Suppose PA $\dashv$ T. If $\tau(x)$ numerates T in a finite subtheory of T, then $T \nvdash \text{Con}_\tau$.

Proof. Let S be a finite subtheory of T and suppose $\tau(x)$ numerates T in S. Let $\tau'(x) := \text{Pr}_S(\tau(\dot{x}))$. Then $\tau'(x)$ is Σ_1 and numerates an extension of T in T. Thus, by Theorem 4(b), $T \nvdash \text{Con}_{\tau'}$. By Corollary 1.8(c), $\text{PA} \vdash \tau'(x) \to \tau(x)$. But then, by Fact 6, $\text{PA} \vdash \text{Con}_\tau \to \text{Con}_{\tau'}$ and so $T \nvdash \text{Con}_\tau$, as desired. ∎

By Theorems 4 and 7, there are formulas $\tau_0(x)$ and $\tau_1(x)$ binumerating T in T such that Con_{τ_0} and Con_{τ_1} are not provably equivalent in T. We now show that this is so even if we restrict ourselves to PR formulas.

Theorem 8. Suppose PA $\dashv$ T. Let $\tau(x)$ be any PR binumeration of T.
  (a) There is a PR binumeration $\tau'(x)$ of T such that
  (i) $T \vdash \text{Con}_\tau \to \text{Con}_{\tau'}$,
  (ii) $T \nvdash \text{Con}_{\tau'} \to \text{Con}_\tau$.
  (b) Let π be a true Π_1 sentence such that $T \vdash \pi \to \text{Con}_\tau$. There is then a PR binumeration $\tau'(x)$ of T such that $T \vdash \pi \leftrightarrow \text{Con}_{\tau'}$.

Proof. (a) Let $\tau'(x)$ be such that
$$\text{PA} \vdash \tau'(x) \leftrightarrow \tau(x) \wedge \forall y \leqslant x \neg \text{Prf}_T(\text{Con}_{\tau'} \to \text{Con}_\tau, y).$$
By Fact 6, (i) holds. Suppose (ii) is false, i.e.
(1) $T \vdash \text{Con}_{\tau'} \to \text{Con}_\tau$.
Let p be a proof of $\text{Con}_{\tau'} \to \text{Con}_\tau$ in T. Then, by Fact 7(a) and Fact 1(v),
$$\text{PA} \vdash \forall y \leqslant x \neg \text{Prf}_T(\text{Con}_{\tau'} \to \text{Con}_\tau, y) \to x < p$$
and so $\text{PA} \vdash \tau'(x) \to \tau(x) \wedge x < p$. By Fact 6, it follows that,
(2) $\text{PA} \vdash \text{Con}_{\tau|p} \to \text{Con}_{\tau'}$.
But $T \vdash \text{Con}_{\tau|p}$, by Corollary 1.8(a). Hence, by (1) and (2), $T \vdash \text{Con}_\tau$, contradicting Theorem 4(a). This proves (ii). Finally, by (ii), Fact 1(iv), and Fact 7(d), $\tau'(x)$ is a PR binumeration of T. ♦

(b) By Fact 5(b), we may assume that $\pi := \forall x \delta(x)$, where $\delta(x)$ is PR. Let $\tau'(x) := \tau(x) \vee \exists y \leqslant x \neg \delta(y)$. Since π is true, $\tau'(x)$ is a PR binumeration of T. Clearly
$$T + \pi \vdash \tau'(x) \to \tau(x).$$
Thus, by Fact 6, $T + \pi \vdash \text{Con}_\tau \to \text{Con}_{\tau'}$ and so $T \vdash \pi \to \text{Con}_{\tau'}$.

To see that the converse implication is provable in T, note that
$$PA \vdash \exists y \neg \delta(y) \to \exists x (\exists y \leqslant x \neg \delta(y) \wedge Pr_\tau(\neg x)).$$
But then, by Corollary 5.4(iv), $PA \vdash \exists y \neg \delta(y) \to \neg Con_{\exists y \leqslant x \neg \delta(y)}$ and so, by Fact 6, $T \vdash Con_{\tau'} \to \pi$. ∎

Suppose $\tau(x)$ is a PR binumeration of T. Then, by Theorem 4, it may be true that $T \vdash \neg Con_\tau$. However, from Theorem 8(a) it follows that we can always choose $\tau(x)$ so that this does not hold:

Corollary 5. If $PA \dashv T$, there is a PR binumeration $\tau(x)$ of T such that $T \nvdash \neg Con_\tau$.

§3. Independent formulas. A formula $\xi(x)$ is *independent over* T if the only propositional combinations of sentences of the form $\xi(k)$ provable in T are the tautologies. This, of course, is the same as saying that $T + \{\xi(k)^{f(k)}: k \in N\}$ is consistent for any $f \in 2^N$.

The following result is a strengthening of Theorem 2.

Theorem 9. There is a Π_1 formula which is independent over T.

Proof. Let $R(k,i,\gamma,p)$ be the primitive recursive relation:
there is a binary sequence s such that $s_k = i$ (so $i = 0$ or $i = 1$) and p is a proof in T of $\neg(\gamma(0)^{s_0} \wedge ... \wedge \gamma(k)^{s_k})$.
Let $\rho(x,y,z,u)$ be a PR binumeration of $R(k,i,\gamma,p)$. Let $\mu(x)$ be such that
$$Q \vdash \mu(x) \leftrightarrow \forall z(\rho(x,1,\mu,z) \to \exists u \leqslant z \rho(x,0,\mu,u)).$$
Suppose, for *reductio ad absurdum*, that $\mu(x)$ is not independent over T. There is then a smallest n for which there is a sequence s such that
(1) $\quad \neg(\mu(0)^{s_0} \wedge ... \wedge \mu(n)^{s_n})$
is provable in T. Let s be the sequence for which the shortest proof p of (1) in T is minimal. There are then two cases. (We assume that $n > 0$ and leave the case $n = 0$ to the reader.)

Case 1. $s_n = 0$. Then
(2) $\quad T \vdash \mu(0)^{s_0} \wedge ... \wedge \mu(n-1)^{s_{n-1}} \to \neg \mu(n)$,
(3) $\quad T \vdash \rho(n,0,\mu,p)$,
(4) $\quad T \vdash \neg \rho(n,1,\mu,q)$ for $q \leqslant p$.
From (3) and (4) we get $T \vdash \mu(n)$ as in the proof of Rosser's theorem. But then, by (2),
(5) $\quad T \vdash \neg(\mu(0)^{s_0} \wedge ... \wedge \mu(n-1)^{s_{n-1}})$,
contrary to the fact that n is minimal.

Case 2. $s_n = 1$. Then
(6) $T \vdash \mu(0)^{s_0} \wedge \ldots \wedge \mu(n-1)^{s_{n-1}} \to \mu(n)$,
(7) $T \vdash \rho(n,1,\mu,p)$,
(8) $T \vdash \neg\rho(n,0,\mu,q)$ for $q < p$.
From (7) and (8) we get $T \vdash \neg\mu(n)$ and so, by (6), we again get (5), again contrary to the minimality of n. ∎

Theorem 9 can be improved as follows; Theorem 10 will be used in Chapter 6 (proof of Lemma 6.8).

Theorem 10. For any Σ_n formula $\delta(x)$, there is a Σ_{n+1} formula $\eta(x)$ such that for any $f, g \in 2^N$, if $T_f = T + \{\delta(k)^{f(k)} : k \in N\}$ is consistent, so is $T_f + \{\eta(k)^{g(k)} : k \in N\}$.

Proof. For every $f \in 2^N$, let $R^f(k,i,\gamma,p)$ be the relation:
    there is a binary sequence s such that $(s)_k = i$ and p is a proof in T_f of
    $\neg(\gamma(0)^{(s)0} \wedge \ldots \wedge \gamma(k)^{(s)k})$
(compare the relation $R(k,i,\gamma,p)$ defined in the proof of Theorem 9). Using the formula $\delta(x)$, we are going to define a formula $\rho^*(x,y,z,w)$ such that for every f,
(1) $\rho^*(x,y,z,w)$ binumerates $R^f(k,i,\gamma,p)$ in T_f.
(Thus, $\rho^*(x,y,z,w)$ behaves in relation to T_f, in the same way as the formula $\rho(x,y,z,u)$ in the proof of Theorem 9 behaves in relation to T.) Let $R^+(k,i,\gamma,t,n,p)$ be the following primitive recursive relation, where t is a binary sequence:
    there is a binary sequence s such that $(s)_k = i$ and p is a proof of
    $\neg(\gamma(0)^{(s)0} \wedge \ldots \wedge \gamma(k)^{(s)k})$ in $T + \delta(0)^{(t)0} + \ldots + \delta(n)^{(t)n}$.
Then
(2) $R^f(k,i,\gamma,p)$ iff $\exists n t \leq p (\forall m \leq n ((t)_m = f(m)) \,\&\, R^+(k,i,\gamma,t,n,p))$.
This is trivial except that it isn't clear that assuming that $R^f(k,i,\gamma,p)$, we can choose $t \leq p$. But this holds if we assume, as we may, that if $\delta(n)^{f(n)}$ occurs in p, then $p \geq 2 \times 3 \times \ldots \times p_n$, where p_n is the nth prime mumber. Let $\rho^+(x,y,z,u,v,w)$ be a PR binumeration of $R^+(k,i,\gamma,t,n,p)$.

By Fact 2, there is a PR formula $\sigma(x,z,u)$ such that
    $Q \vdash \sigma(k,m,u) \leftrightarrow u = (k)_m$.
Let
    $\beta(x,y) := \forall z \leq y ((\delta(z) \to \sigma(x,z,0)) \wedge (\neg\delta(z) \to \sigma(x,z,1)))$.
Then for every n and every t,
(3) $T_f \vdash \beta(t,n) \leftrightarrow \bigwedge\{(t)_m = f(m) : m \leq n\}$.
In view of (2) and (3), the obvious definition of $\rho^*(x,y,z,w)$ is now:

§3. *Independent Formulas*   37

$$\rho^*(x,y,z,w) := \exists uv \leq w(\beta(u,v) \wedge \rho^+(x,y,z,u,v,w)).$$

To prove (1), suppose first $R^f(k,i,\gamma,p)$. By (2), there are then n, t ⩽ p such that $(t)_m = f(m)$ for all m ⩽ n and $R^+(k,i,\gamma,t,n,p)$. But then $T \vdash \rho^+(k,i,\gamma,t,n,p)$. By (3), it follows that $T_f \vdash \beta(t,n)$ and so that $T_f \vdash \rho^*(k,i,\gamma,p)$.

Next suppose $\neg R^f(k,i,\gamma,p)$. Then, by (2), $\neg R^+(k,i,\gamma,t,n,p)$ for every n ⩽ p and every t ⩽ p such that $(t)_m = f(m)$ for all m ⩽ n. It follows that $T \vdash \neg \rho^+(k,i,\gamma,t,n,p)$ for all such n and t. Also, by (3), $T_f \vdash \neg \beta(t,n)$ for all t such that $(t)_m \neq f(m)$ for some m ⩽ n. It follows that $T_f \vdash \neg \rho^*(k,i,\gamma,p)$. This proves (1).

Let $\eta(x)$ be such that
$$Q \vdash \eta(x) \leftrightarrow \exists z(\rho^*(x,0,\eta,z) \wedge \forall u \leq z \neg \rho^*(x,1,\eta,u)).$$
The proof that $\eta(x)$ is as desired is now the same as the proof of Theorem 9 except that T is replaced by any consistent theory T_f, and the fact that $\rho(x,y,z,u)$ is decidable in T is replaced by (1). We leave this part of the proof to the reader.

Finally, if $\delta(x)$ is Σ_n, then $\beta(x,y)$ is Δ_{n+1}, whence the same is true of $\rho^*(x,y,z,w)$ and so $\eta(x)$ is Σ_{n+1}, as desired. ∎

The proof of the final theorem of this § is quite different from the proofs of Theorems 9 and 10; instead of a Rosser type construction it uses the formulas $\text{Sat}_\Phi(x,y)$ and so does not apply to Q (and its extensions).

In the proof of Theorem 11 we assume, as we may, that
(+) $PA \vdash \langle x,y \rangle = z \leftrightarrow (x = (z)_0 \wedge y = (z)_1).$

Theorem 11. Suppose $PA \dashv T$. Then there is a Γ (Δ_{n+1}) formula $\gamma(x)$ such that $T + \forall x(\gamma(x) \leftrightarrow \delta(x))$ is consistent for every Γ (B_n) formula $\delta(x)$.

Proof. Suppose $\Gamma = \Sigma_n$; the case $\Gamma = \Pi_n$ follows by taking negations. Let $S(k,m,n)$ be a primitive recursive relation such that
$$\delta(x) \in \Sigma_n \;\&\; T \vdash \neg \forall x(\eta(x) \leftrightarrow \delta(x)) \text{ iff } \exists n S(\eta,\delta,n).$$
Let $\sigma(x,y,z)$ be a PR binumeration of $S(k,m,n)$ and let $\sigma^*(x,y,z) :=$
$$\sigma(x,y,z) \wedge \forall y'z'(\langle y',z' \rangle < \langle y,z \rangle \to \neg \sigma(x,y',z')).$$
Finally, let $\gamma(x)$ be such that
(1) $PA \vdash \gamma(x) \leftrightarrow \exists yz(\sigma^*(\gamma,y,z) \wedge \text{Sat}_{\Sigma_n}(x,y)).$
Suppose there is a Σ_n formula $\delta(x)$ such that
(2) $T \vdash \neg \forall x(\gamma(x) \leftrightarrow \delta(x)).$
For each such formula, there is an n such that $S(\gamma,\delta,n)$. Now pick $\delta(x)$ and n so that $\langle \delta,n \rangle$ is minimal. Then, by (+),
$$PA \vdash \sigma^*(\gamma,y,z) \leftrightarrow y = \delta \wedge z = n.$$

Hence, by (1) and Fact 10(a)(i), $PA \vdash \forall x(\gamma(x) \leftrightarrow \delta(x))$, contradicting (2). Thus, (2) is false for all Σ_n formulas $\delta(x)$, as desired.

To obtain a Δ_{n+1} formula as desired, replace $Sat_{\Sigma_n}(x,y)$ by $Sat_{B_n}(x,y)$ in (1). ∎

For extensions T of PA, Theorem 9 follows at once from Theorem 11.

Clearly, the Γ (Δ_{n+1}) formula of Theorem 11 is not $\Gamma^{d,T}$ (B_n^T) (compare Theorem 1.4). Theorem 11 also has the following:

Corollary 6. Suppose $PA \dashv T$. There is a Γ (Δ_{n+1}) sentence not in $\Gamma^{d,T}$ (B_n^T).

Proof. Let $\gamma(x)$ be as in Theorem 11 and let $\varphi := \gamma(0)$. ∎

§4. The length of proofs. We begin by showing that the length of proofs of (Π_1) sentences φ is not bounded by any recursive function of φ.

Theorem 12. Let f be any recursive function. There is then a sentence φ such that $T \vdash \varphi$ and the least proof of φ in T is $> f(\varphi)$.

Proof. Let $\delta_f(x,y)$ be a Σ_1 formula defining f in Q (cf. Fact 3(b)). Let φ be such that
$$Q \vdash \varphi \leftrightarrow \forall y(\delta_f(\varphi,y) \rightarrow \forall z \leqslant y \neg Prf_T(\varphi,z)).$$
Suppose φ has a proof $p \leqslant f(\varphi)$ in T. Since
$$Q \vdash \delta_f(\varphi,y) \leftrightarrow y = f(\varphi)$$
and, by Fact 7(a), $Q \vdash Prf_T(\varphi,p)$, it follows that $Q \vdash \neg \varphi$ and so $T \vdash \neg \varphi$, a contradiction. Thus, φ has no proof $p \leqslant f(\varphi)$ in T. But then, by Fact 1(iv) and Fact 7(d), $Q \vdash \forall z \leqslant f(\varphi) \neg Prf_T(\varphi,z))$, whence $Q \vdash \varphi$ and so $T \vdash \varphi$. ∎

In Theorem 12 and in Theorems 13 and 14, below, we use (the Gödel number of) the proof as a measure of its "length". We could also have used the number of (occurrences of) symbols as a (more natural) measure of "length" and proved the same results.

Suppose $T \nvdash \varphi$. Then $T + \varphi$ is stronger than T not only in the sense that it proves more theorems but also in the sense that there are infinitely many theorems of T which have "much shorter" proofs in $T + \varphi$.

Theorem 13. Suppose $T \nvdash \varphi$. Let f be any recursive function. There is then a sentence θ such that $T \vdash \theta$ and there is a proof q of θ in $T + \varphi$ such that θ has no proof $\leqslant f(q)$ in T.

§4. The Length of Proofs

Proof. Suppose not and let f be a counterexample. $T + \neg\varphi$ is consistent. Let g be a recursive function such that for any sentence ψ, $g(\psi)$ is a proof of $\varphi \vee \psi$ in $T + \varphi$. Then, by assumption, $T + \neg\varphi \vdash \psi$ iff $T \vdash \varphi \vee \psi$ iff $\varphi \vee \psi$ has a proof $\leq f(g(\psi))$ in T. It follows that $T + \neg\varphi$ is decidable, contradicting Theorem 1.2. ∎

Another way of obtaining "much shorter proofs", in this case without getting any new theorems, is to add new (nonlogical but correct) rules of inference. For example, if T is Σ_1-sound, the rule

R: from $\Pr_T(\varphi)$ derive φ,

is *correct for* T in the sense that every sentence which can be derived (from the axioms of T) using this rule can be proved without it, i.e., is a theorem of T. That R occasionally leads to "much shorter proofs" follows from:

Theorem 14. Suppose $PA \dashv T$ and T is Σ_1-sound. Let $g(k,m)$ be any primitive recursive function. There are then a (Σ_1, Π_1) sentence φ such that $T \vdash \varphi$ and a proof q of $\Pr_T(\varphi)$ in T such that φ has no proof $\leq g(\varphi,q)$ in T.

Proof. We may assume that $g(k,m)$ is increasing in m. Let φ be such that
$$T \vdash \varphi \leftrightarrow \exists y(\mathrm{Prf}_T(\Pr_T(\varphi),y) \wedge \forall z \leq g(\varphi,y) \neg \mathrm{Prf}_T(\varphi,z)).$$
Clearly
$$T + \Pr_T(\Pr_T(\varphi)) + \neg \Pr_T(\varphi) \vdash \varphi.$$
Since φ is Σ_1, we have, by provable Σ_1-completeness, $T + \varphi \vdash \Pr_T(\varphi)$. It follows that
$$T + \Pr_T(\Pr_T(\varphi)) \vdash \Pr_T(\varphi),$$
and so, by Theorem 6, $T \vdash \Pr_T(\varphi)$. Since T is Σ_1-sound, this implies that $T \vdash \varphi$ and that φ is true.

Let q be the least proof of $\Pr_T(\varphi)$ in T. Since φ is true and $g(k,m)$ is increasing in m, it follows that φ has no proof $\leq g(\varphi,q)$.

To obtain a Π_1 sentence as desired, let φ be such that
$$T \vdash \varphi \leftrightarrow \forall z(\mathrm{Prf}_T(\varphi,z) \rightarrow \exists y \leq z(g(\varphi,y) < z \wedge \mathrm{Prf}_T(\Pr_T(\varphi),y)))$$
and set
$$\varphi^* := \exists y(\mathrm{Prf}_T(\Pr_T(\varphi),y) \wedge \forall z \leq g(\varphi,y) \neg \mathrm{Prf}_T(\varphi,z)).$$
Then
$$T + \Pr_T(\Pr_T(\varphi)) + \neg \Pr_T(\varphi) \vdash \varphi^*.$$
Clearly, $T \vdash \varphi^* \rightarrow \varphi$ and so, by (BLi) and (BLii), $T \vdash \Pr_T(\varphi^*) \rightarrow \Pr_T(\varphi)$. Since φ^* is Σ_1, we have $T + \varphi^* \vdash \Pr_T(\varphi^*)$. It follows that
$$T + \Pr_T(\Pr_T(\varphi)) \vdash \Pr_T(\varphi).$$
The rest of the proof is now the same as above, except that we observe that, since φ is Π_1 and $T \vdash \varphi$, φ must be true (Fact 9(a)). ∎

For any sequence p of formulas and any formula θ, let p^θ be p followed by θ. If p is a proof of $\Pr_T(\theta)$ in T we may think of p^θ as an R-proof of θ in T, i.e., a proof in T in which we are allowed to use the rule R. Now, let h be any primitive recursive function and let $g(\theta,p) = h(p^\wedge\theta)$. Then g is primitive recursive. Let φ and q be as in Theorem 14 and let $r = q^\wedge\varphi$. Then r is an R-proof of φ in T and φ has no proof $\leq h(r)$ in T.

Exercises for Chapter 2

In the following Exercises we write Prf(x,y), Pr(x), Con for $\Prf_T(x,y)$, $\Pr_T(x)$, $\Con_T$, respectively.

1. Suppose U is a (not necessarily r.e.) extension of Q.
    (a) Improve Lemma 2 by showing that there is no formula numerating $N - \Th(U)$ in U.
    (b) Suppose U is true and there is a formula binumerating U in U. Show that U is not complete.

2. (a) Suppose T is Σ_1-sound. Use the fact that there is an r.e. nonrecursive set to show that there is a (true) Π_1 sentence not provable in T.
    (b) Let $\Ref(T) = \{\varphi: T \vdash \neg\varphi\}$. Let X be any set such that $\Th(T) \subseteq X$ and $\Ref(T) \cap X = \emptyset$. Show that there is no formula binumerating X in T. (This improves Lemma 1.2.) Conclude that $\Th(T)$ and $\Ref(T)$ are *recursively inseparable*, i.e., there is no recursive set Y such that $\Th(T) \subseteq Y$ and $\Ref(T) \cap Y = \emptyset$. (This implies Theorem 1.2.)

3. (a) Let X_0 and X_1 be disjoint r.e. sets. Let $\rho_i(x,y)$ be a PR formula such that $X_i = \{k: \exists m Q \vdash \rho_i(k,m)\}$, $i = 0, 1$. Let
    $$\xi(x) := \exists y(\rho_0(x,y) \wedge \forall z \leq y \neg \rho_1(x,z)).$$
    Show that if $k \in X_0$, then $Q \vdash \xi(k)$, and if $k \in X_1$, then $Q \vdash \neg\xi(k)$.
    (b) Use the fact that there are recursively inseparable r.e. sets to show that there is a (true) Π_1 sentence which is undecidable in T (compare Theorem 2).
    (c) Show that the sets of Π_1 and Σ_1 sentences provable in T are not recursive.

4. Suppose PA $\dashv$ T. Show that the set of Δ_n^T formulas is not recursive. [Hint: Let $\sigma(x)$ be a Σ_n formula which is not Δ_n^T. Let X_i and $\rho_i(x,y)$ be as in Exer-

cise 3(a). Suppose X_0 and X_1 are recursively inseparable. Let $\eta(x,y) :=$
$\sigma(x) \vee \exists z(\rho_0(y,z) \wedge \forall u \leqslant z \neg \rho_1(y,u))$.
Let $Y = \{k: \eta(x,k) \text{ is } \Delta_n^T\}$. Then $X_0 \subseteq Y$ and $X_1 \cap Y = \emptyset$.]

5. (a) Let Y be any r.e. set of formulas $\eta(x)$ all of which are decidable in T. Show that there is a recursive set X such that no member of Y binumerates X in T.

(b) Suppose PA ⊣ T. Show that every Δ_1 formula is decidable in T. Conclude that there is a recursive set which is not binumerated by a Δ_1 formula in T (compare Corollary 1.3(a)).

(c) A formula $\xi(x)$ is *provably decidable* in T if $T \vdash \forall x(\text{Pr}(\xi(\dot{x})) \vee \text{Pr}(\neg \xi(\dot{x})))$. (PR formulas are provably decidable in PA.) Suppose T is Σ_1-sound. Show that there is a recursive set which is not binumerated in T by a formula which is provably decidable in T. Conclude that there is a (Σ_1) formula which is decidable in Q but not provably decidable in T.

6. Suppose PA ⊣ T.
(a) Show that the formula $\text{Tr}_\Gamma(x)$ is (Γ but) not $\Gamma^{d,T}$.
(b) Show that the formula $\text{Tr}_{B_n}(x)$ is (Δ_{n+1} but) not B_n^T. (Compare Theorem 1.4).

7. (a) Suppose T is Σ_1-sound. Show that not every recursive function is provably recursive in T (compare Exercise 36(b)). [Hint: Let $\delta_0(x,y), \delta_1(x,y),\ldots$ be an effective enumeration of all Σ_1 formulas $\delta(x,y)$ provably in T defining (total) functions, i.e., such that
(tot) $T \vdash \forall x \exists y \forall z(\delta(x,z) \leftrightarrow z = y)$.
For each m, let $f_m(k)$ be the recursive function defined by $\delta_m(x,y)$ in T. The function $g(k) = f_k(k)+1$ is then recursive.]

(b) Show that for each recursive function f(k), there is a formula $\delta(x,y)$ defining f in T and such that (tot) holds. (Thus, the restriction to Σ_1 formulas in the definition of "provably recursive" is essential.)

(c) Suppose PA ⊣ T. Show that if T is not Σ_1-sound, then every recursive function is provably recursive in T. (Thus, the restriction to Σ_1-sound theories in (a) is essential.)

8. (a) Let φ, ψ be Γ sentences such that $T \vdash \varphi \to \psi$ and $T \nvdash \psi \to \varphi$. Show that there is then a Γ sentence θ such that $T \vdash \varphi \to \theta$, $T \vdash \theta \to \psi$, $T \nvdash \theta \to \varphi$, and $T \nvdash \psi \to \theta$.

(b) Suppose Q ⊣ S. Show that there is a Π_1 sentence θ such that $S \nvdash \theta$,

$S \nvdash \neg\theta$, $T \nvdash \neg Pr_S(\theta)$, $T \nvdash \neg Pr_S(\neg\theta)$.

9. Show that if π is a Π_1 sentence and $T \vdash \pi$, then $PA + \text{Con} \vdash \pi$. [Hint: Use Fact 9.]

10. Suppose $PA \dashv T$. Suppose φ is undecidable in T. Show that there is a (Σ_1) sentence ψ such that
$$T \nvdash \varphi \to \psi,$$
$$PA \vdash Pr(\varphi) \to Pr(\psi).$$
[Hint: Let ψ be such that
$$Q \vdash \psi \leftrightarrow \exists y (Prf(\varphi,y) \wedge \forall z \leqslant y \neg Prf(\varphi \to \psi, z)).$$
Then $PA \vdash Pr(\varphi) \wedge \neg Pr(\varphi \to \psi) \to Pr(\psi)$.]

11. Suppose $\alpha(x)$ is such that for every sentence φ, $\alpha(\varphi)$ "says" that φ is (in some sense) a necessary truth. One would then expect that for T strong enough and for all φ, ψ,
(i) $T \vdash \alpha(\varphi) \to \varphi$,
(ii) $T \vdash \alpha(\alpha(\varphi) \to \varphi)$,
(iii) if $\vdash \varphi$, then $T \vdash \alpha(\varphi)$,
(iv) $T \vdash \alpha(\varphi)$ and $T \vdash \alpha(\varphi \to \psi)$, then $T \vdash \alpha(\psi)$.
Show that there is no formula $\alpha(x)$ satisfying these conditions.
[Hint: Let $\chi := \wedge Q$. Let φ be such that
$$Q \vdash \varphi \leftrightarrow \neg \alpha(\chi \to \varphi).$$
It follows that
$$\vdash (\alpha(\chi \to \varphi) \to (\chi \to \varphi)) \to (\chi \to \varphi).$$
Show that $T \vdash \varphi$ and $T \vdash \neg \varphi$.]

12. Suppose $PA \dashv T$.
   (a) Let φ be as in (G). Show that if T is Σ_1-sound, then $T \nvdash \text{Con} \to \neg Pr(\neg\varphi)$ (compare Theorem 5).
   (b) Let θ be a Π_1 Rosser sentence for T and let $\psi :=$
$$\forall u (Prf(\neg\theta, u) \to \exists z < u Prf(\theta, z)).$$
Show that $T \nvdash \theta \to \text{Con}$, $T \nvdash \psi \to \text{Con}$, and $PA \vdash \theta \wedge \psi \to \text{Con}$. Conclude that $T \nvdash \psi$. [Hint: Use Theorem 4 and Corollary 3.]

13. Suppose $PA \dashv T$. Let f be any recursive function. Show that there is a Π_1 sentence θ such that $T \vdash \theta$ and $T | f(\theta) \nvdash \theta$. (This improves Corollary 1; also compare Exercise 4.5.) [Hint: Let $\delta_f(x,y)$ be a Σ_1 formula defining f in Q (cf. Fact 3(b)) and let θ be such that

$$Q \vdash \theta \leftrightarrow \forall y (\delta_f(\theta,y) \to \neg \mathrm{Pr}_{T|y}(\theta)).]$$

14. Suppose PA ⊣ T.
   (a) Prove Löb's theorem by considering a sentence θ such that
   (*) $\mathrm{PA} \vdash \chi \leftrightarrow \mathrm{Pr}(\chi \to \varphi)$.
   (This is essentially the proof of Theorem 6 using Theorem 4 mentioned in the text.) Show that this proof, too, can be formalized in PA.
   (b) Show that (i) if θ is as in (L), then $\chi := \mathrm{Pr}(\theta)$ is as in (*) and that (ii) if χ is as in (*), then $\theta := \mathrm{Pr}(\chi) \to \chi$ is as in (L).

15. Let φ be as in Theorem 1.
   (a) Show that $\mathrm{PA} \vdash \varphi \to \mathrm{Con}$. Conclude that $\mathrm{PA} \vdash \varphi \leftrightarrow \mathrm{Con}$ and so $\mathrm{PA} \vdash \mathrm{Con} \leftrightarrow \neg \mathrm{Pr}(\mathrm{Con})$.
   (b) Show that (L) (in the proof of Theorem 6) is true if $\theta := \mathrm{Pr}(\varphi) \to \varphi$. (Thus, there are sentences φ and θ, namely Con and $\mathrm{Pr}(\varphi) \to \varphi$, satisfying (G) with Q replaced by PA and (L), respectively, and not constructed using self-reference.)

16. Suppose PA ⊣ T. Show that there is a PR formula $\delta(x)$ such that $T \vdash \forall x \mathrm{Pr}(\delta(\dot{x}))$ and $T \nvdash \forall x \delta(x)$. [Hint: Let φ be as in the proof of Theorem 1 and let $\delta(x) := \neg \mathrm{Prf}(\varphi,x)$. Then $\mathrm{PA} \vdash \neg \delta(x) \to \mathrm{Pr}(\bot)$.]

17. Suppose PA ⊣ T. Let $\tau(x)$ be a PR binumeration of T and let $\tau^*(x)$ be as in Theorem 7.
   (a) Let ψ be such that $\mathrm{PA} \vdash \psi \leftrightarrow \neg \mathrm{Pr}_{\tau^*}(\psi)$. Show that ψ is undecidable in T. [Hint: For every sentence φ, $\mathrm{PA} \vdash \mathrm{Pr}_{\tau^*}(\varphi) \to \neg \mathrm{Pr}_{\tau^*}(\neg \varphi)$.]
   (b) Show that $T \vdash \mathrm{Con}_{\tau^* + \neg \mathrm{Con}_\tau}$. [Hint: Let φ be as in Theorem 1. Then $T \vdash \neg \varphi \to \mathrm{Pr}_{\tau^*}(\neg \varphi)$ and so $T \vdash \neg \varphi \to \neg \mathrm{Pr}_{\tau^*}(\varphi)$. It follows that $T \vdash \mathrm{Pr}_\tau(\varphi) \to \neg \mathrm{Pr}_{\tau^*}(\varphi)$. Also $T \vdash \neg \mathrm{Pr}_\tau(\varphi) \to \neg \mathrm{Pr}_{\tau^*}(\varphi)$ and so $T \vdash \neg \mathrm{Pr}_{\tau^*}(\varphi)$. Now use the fact that $T \vdash \mathrm{Con}_\tau \to \varphi$.]

18. Suppose PA ⊣ T. Prove the following strengthening of Corollary 5. Suppose X is r.e. and monoconsistent with T. There is then a PR binumeration $\tau(x)$ of T such that $\neg \mathrm{Con}_\tau \notin X$ (see Exercise 6.6(b)). [Hint: Let $\tau'(x)$ be a PR binumeration of T and let $\rho(x,y)$ be a PR binumeration of a relation $R(k,m)$ such that $X = \{k: \exists m R(k,m)\}$. Let $\tau(x) := \tau'(x) \wedge \forall y \leq x \neg \rho(\neg \varphi, y)$, where φ is such that $\mathrm{PA} \vdash \varphi \leftrightarrow \mathrm{Con}_\tau$.]

19. Suppose PA ⊣ T. Let $\tau_0(x)$ and $\tau_1(x)$ be PR binumerations of T.

(a) Show that there is a PR binumeration $\tau(x)$ of T such that
$$T \vdash \mathrm{Con}_\tau \leftrightarrow \mathrm{Con}_{\tau_0} \wedge \mathrm{Con}_{\tau_1}.$$
[Hint: Let $\tau(x) := \tau_0(x) \vee \exists y \leqslant x \mathrm{Prf}_{\tau_1}(\bot, y)$. See also Theorem 8(b).]

(b) Show that there is a PR binumeration $\tau(x)$ of T such that
$$T \vdash \mathrm{Con}_\tau \leftrightarrow \mathrm{Con}_{\tau_0} \vee \mathrm{Con}_{\tau_1}.$$
[Hint: Let $\tau(x) := (\tau_0(x) \wedge \tau_1(x)) \vee (\exists y \leqslant x \mathrm{Prf}_{\tau_0}(\bot, y) \wedge \exists y \leqslant x \mathrm{Prf}_{\tau_1}(\bot, y))$.]

(c) Suppose $T \nvdash \varphi$. Show that there is a PR binumeration $\tau(x)$ of T such that $T \nvdash \mathrm{Con}_\tau \to \varphi$ (compare Theorem 8(a)).

(d) Suppose $T \nvdash \varphi$ and $T \nvdash \neg \psi$. Show that there is a PR binumeration $\tau(x)$ of T such that $T \nvdash \mathrm{Con}_\tau \to \varphi$ and $T \nvdash \psi \to \mathrm{Con}_\tau$. [Hint: Use Lemma 1 and Theorem 8(b).]

20. Strengthen Lemma 1 in the following way. Suppose X is r.e. and monoconsistent with Q. Show that there is a Π_1 formula $\eta(x)$ such that the only propositional combinations of sentences of the form $\eta(k)$ which are members of X are the tautologies.

21. Prove the following strengthening of Theorems 3 and 9. If $\{T_k : k \in \mathbb{N}\}$ is an r.e. family of theories, there is a Π_1 formula which is simultaneously independent over all the theories T_k. Strengthen Theorems 10 and 11 in the same way.

22. (a) Derive Theorem 9 for extensions of PA from Theorem 11.

(b) Formulate and prove a generalization of Theorem 11 which implies Theorem 10 for extensions of PA.

(c) Improve Theorem 11 as follows. Suppose $PA \dashv T$. Let X be an r.e. set monoconsistent with T. Show that there is a Γ (Δ_{n+1}) formula $\gamma(x)$ such that $\neg \forall x (\gamma(x) \leftrightarrow \delta(x)) \notin X$ for every Γ (B_n) formula $\delta(x)$.

23. Derive Theorem 12 from Theorem 1.2.

24. Suppose $PA \dashv T$. Show that there is a Π_1 formula $\kappa(x)$ such that $T \nvdash \kappa(k)$ for every k, but $T \vdash \kappa(k) \vee \kappa(m)$ whenever $k \neq m$. (This can also be obtained as a special case of Theorem 3.5.). [Hint: Let $\kappa(x)$ be such that
$$PA \vdash \kappa(x) \leftrightarrow \forall y (\mathrm{Prf}(\kappa(\dot{x}), y) \to \exists z u (\langle z, u \rangle < \langle x, y \rangle \wedge \mathrm{Prf}(\kappa(\dot{z}), u))).]$$

25. Suppose $PA \dashv T$. Show that the following conditions are equivalent:
(i) T is Σ_1-sound.
(ii) For any two Σ_1 sentences σ_0, σ_1, if $T \vdash \sigma_0 \vee \sigma_1$, then either $T \vdash \sigma_0$ or

$T \vdash \sigma_1$.

(iii) If φ is Δ_1^T, then either $T \vdash \varphi$ or $T \vdash \neg \varphi$.
(iv) $Pr(x)$ numerates $Th(T)$ in T (compare Exercise 6.17).
[Hint: (ii) implies (i). Let $\delta(z)$ be a PR formula such that $\exists z \delta(z)$ is false and provable in T. Let σ_0, σ_1 be such that
$$Q \vdash \sigma_0 \leftrightarrow \exists z((Prf(\neg \sigma_0, z) \vee \delta(z)) \wedge \forall u \leqslant z \neg Prf(\sigma_0, u)),$$
$$\sigma_1 := \exists u(Prf(\sigma_0, u) \wedge \forall z \leqslant u(\neg Prf(\neg \sigma_0, z) \wedge \neg \delta(z))).]$$

26. Suppose PA ⊣ T, S.

(a) Let σ be any Σ_1 sentence such that $T \vdash Pr_S(\bot) \to \sigma$. Show that there is a Σ_1 sentence χ such that $T \vdash \sigma \leftrightarrow Pr_S(\chi)$. Conclude that (i) for every sentence φ, there is a Σ_1 sentence χ such that $T \vdash Pr_S(\chi) \leftrightarrow Pr_S(\varphi)$, (ii) for any sentences φ, ψ, there is a Σ_1 sentence χ such that $T \vdash Pr_S(\chi) \leftrightarrow Pr_S(\varphi) \vee Pr_S(\psi)$, (iii) for every Π_1 sentence π such that $T \vdash \pi \to Con_S$, there is a Π_1 sentence θ such that $T \vdash \pi \leftrightarrow Con_{S+\theta}$; in particular, if S' is any theory and $T \vdash Con_{S'} \to Con_S$, then there is a ($\Pi_1$) sentence θ such that $T \vdash Con_{S'} \leftrightarrow Con_{S+\theta}$ (compare Theorem 8(b)). [Hint: Let $\delta(y)$ be a PR formula such that $\sigma := \exists y \delta(y)$. Let χ be such that
$$PA \vdash \chi \leftrightarrow \exists y(\delta(y) \wedge \forall z \leqslant y \neg Prf_S(\chi, z))$$
and let $\chi^* := \exists z(Prf_S(\chi, z) \wedge \forall y < z \neg \delta(y))$. Then χ^* is Σ_1, $PA \vdash \chi^* \to \neg \chi$, and $PA \vdash Pr_S(\chi) \wedge \neg \sigma \to \chi^*$.]

(b) Improve (a) as follows. Let σ_i, $i = 0, 1$, be any Σ_1 sentences. Show that $T \vdash \sigma_0 \wedge \sigma_1 \leftrightarrow Pr_S(\bot)$ iff there is a (Σ_1) sentence χ such that $T \vdash \sigma_0 \leftrightarrow Pr_S(\chi)$ and $T \vdash \sigma_1 \leftrightarrow Pr_S(\neg \chi)$.

27. Suppose PA ⊣ T. Let $\tau(x)$ be a PR binumeration of T. Let $\alpha(x), \beta(x)$ be any formulas and let $\alpha \leqslant \beta$ mean that there is a primitive recursive function g such that
$$PA \vdash \forall x(Prf_\alpha(\bot, x) \to Prf_\beta(\bot, g(x))).$$
Thus, $\alpha \leqslant \beta$ implies $PA \vdash Con_\beta \to Con_\alpha$. (Most real-life proofs of statements of the form $Con_\beta \to Con_\alpha$ (relative consistency proofs) actually yield the stronger conclusion that $\alpha \leqslant \beta$.)

(a) Show that $\tau + \neg Con_\tau \leqslant \tau$.

(b) Suppose $PA \vdash \tau(x) \to \alpha(x)$. Show that there is a Π_1 (Σ_1) sentence φ such that $\alpha \leqslant \tau + \varphi \leqslant \alpha$. [Hint: In the Π_1 case let φ be such that
$$PA \vdash \varphi \leftrightarrow \forall x(Prf_\alpha(\bot, x) \to \exists y \leqslant x Prf_{\tau + \varphi}(\bot, y)).$$
It follows from Fact 9(c) that for every sentence ψ and every primitive recursive function f, there is a primitive recursive function g such that
$$PA \vdash \forall x(Prf_\alpha(\exists y \leqslant \dot{x}\, Prf_\alpha(\psi, f(y))) \to \psi, g(x))).]$$

(c) There is a function f (for example, the Ackermann function) such that the following holds. f is provably recursive in PA. Let $\delta_f(x,y)$ be a Σ_1 formula defining f in PA and such that (tot) holds with T = PA (see Exercise 7). For every primitive recursive function h, there is an n such that $PA \vdash \forall x > n \exists y(\delta_f(x,y) \land h(x) < y)$. Use this to show that there is a (Σ_1) sentence φ such that $PA \vdash Con_\tau \to Con_{\tau+\varphi}$ and $\tau + \varphi \not\geq \tau$. [Hint: Let φ be such that
$$PA \vdash \varphi \leftrightarrow \exists y(Prf_{\tau+\varphi}(\bot,y) \land \exists z(\delta_f(y,z) \land \forall u \leq z \neg Prf_\tau(\bot,u))).$$
First show that $PA \vdash Con_\tau \to Con_{\tau+\varphi}$ and then that if $\tau + \varphi \leq \tau$, then $PA \vdash \neg \varphi$.]

(d) Let φ be any sentence such that T + φ is consistent. Show that there is a PR binumeration $\tau'(x)$ of T such that $\tau' + \varphi \leq \tau'$ (compare Theorem 8(b)).

28. Suppose PA ⊣ T.

(a) Let φ be any Γ sentence. Show that there is a formula $\xi(x)$ such that if ψ is Γ, then ψ is a fixed point of $\xi(x)$ in T iff $\psi := \varphi$.

(b) Suppose $\gamma(x)$ is Γ. Show that $\gamma(x)$ has infinitely many Γ fixed points in T. Conclude that the formula $\xi(x)$ mentioned in (a) cannot be Γ.

(c) Let X be any r.e. set of sentences. Show that there is a formula $\xi(x)$ such that if $\varphi \in X \cap \Gamma$, then φ is a fixed point of $\xi(x)$ in T and if $\varphi \in \Gamma - X$, then φ is not a fixed point of $\xi(x)$ in T. [Hint: Let $\rho(x,y)$ be a PR formula such that $X = \{k: \exists m \, PA \vdash \rho(k,m)\}$. Let $\xi(x)$ be such that
$$PA \vdash \xi(\varphi) \leftrightarrow (Tr_\Gamma(\varphi) \land \exists y(\rho(\varphi,y) \land \forall z \leq y \neg Prf(\varphi \leftrightarrow \xi(\varphi),z))) \lor$$
$$(\neg Tr_\Gamma(\varphi) \land \exists z(Prf(\varphi \leftrightarrow \xi(\varphi),z) \land \forall y \leq z \neg \rho(\varphi,y))).]$$

29. (a) Show that if T is Σ_n-sound, then T is Π_{n+1}-sound.

(b) Suppose PA ⊣ T and T is Σ_{n+1}-sound. Let $\sigma_n :=$
$$\forall z(\Sigma_n(z) \land Pr(z) \to Tr_{\Sigma_n}(z)).$$
Thus, σ_n "says" that T is Σ_n-sound. (These sentences will reappear in Chapter 4.) Show that $T \not\vdash \sigma_n \to \sigma_{n+1}$. [Hint: $T + \sigma_{n+1} \vdash Con_{T+\sigma_n}$.]

(c) Suppose PA ⊣ T and T is Σ_n-sound. Let φ be such that
$$PA \vdash \varphi \leftrightarrow \exists z(\Sigma_n(z) \land Pr_{T+\varphi}(z) \land \neg Tr_{\Sigma_n}(z)).$$
(These sentences will reappear in Chapter 5.) Show that φ is false. Conclude that Σ_n-soundness does not imply Σ_{n+1}-soundness.

30. T is ω-*consistent* iff for every formula $\alpha(x)$, if $T \vdash \neg \alpha(k)$ for every k, then $T \not\vdash \exists x \alpha(x)$.

(a) Show that if T is ω-consistent, then T is Π_3-sound.

(b) Suppose T is true. Show that there is a false Σ_3 sentence φ such that

T + φ is ω-consistent. Conclude that ω-consistency does not imply Σ_3-soundness. [Hint: Let φ be a sentence "saying" that T + φ is not ω-consistent.]

(c) Suppose PA ⊣ T and T is true. Show that for every n, there is an extension S of T which is Σ_n-sound but not ω-consistent. [Hint: Let δ(x) be a Π_n formula such that PA ⊢ φ ↔ ∃xδ(x), where φ as in Exercise 29(c). Let S = T + ∃xδ(x) + {¬δ(k): k∈N}.]

31. Let T, φ, f be as in Theorem 13.
  (a) Suppose φ is Γ. Show that θ in the theorem can be taken to be Γ.
  (b) Construct a sentence θ satisfying the conclusion of the theorem.

In Exercises 32–36 "proof" means "proof in T". Let
  $Prf^\mu(x,y) := Prf(x,y) \land \forall z{<}y \neg Prf(x,z)$.

32. Let f be any recursive function.
  (a) The sentence φ defined in he proof of Theorem 12 is Π_1. Show that there is a Σ_1 sentence φ such that T ⊢ φ and the least proof of φ is > f(φ).
  (b) Show that there is a Π_1 formula ξ(x) such that for every n, T ⊢ ξ(n) and the least proof of ξ(n) is > f(n).

33. Suppose PA ⊣ T and let g be any recursive function. Show that there are Π_1 sentences ψ_0, ψ_1 provable in T and a proof p of $\psi_0 \lor \psi_1$ such that neither ψ_0 nor ψ_1 has a proof ⩽ g(p). [Hint: Let $\delta_g(x,y)$ be a Σ_1 formula defining g in T. Let ψ_i be such that
  T ⊢ ψ_i ↔ ∀yz($Prf^\mu(\psi_0 \lor \psi_1, y) \land$
      $\exists v(\delta_g(y,v) \land z \leqslant v) \land Prf(\psi_i, z) \rightarrow \exists u{<}z{+}i\, Prf(\psi_{1-i}, u))$.]

34. Suppose PA ⊣ T and T is Σ_1-sound. There is then a recursive function g which given a proof p of a sentence $Pr(\phi_0) \lor Pr(\phi_1)$ picks out a ϕ_i such that $Pr(\phi_i)$ is true; in other words, g(p) = 0 or g(p) = 1, if g(p) = 0, then $Pr(\phi_0)$ is true, and if g(p) = 1, then $Pr(\phi_1)$ is true. Show that g is not primitive recursive even if we restrict ourselves to Σ_1 sentences ϕ_0, ϕ_1. [Hint: Suppose not. Assume that T ⊢ g(y) = 0 ∨ g(y) = 1. Let ψ_i be such that
  T ⊢ ψ_i ↔ ∃y($Prf^\mu(Pr(\psi_0) \lor Pr(\psi_1), y) \land g(y) = 1-i$).]

35. Suppose PA ⊣ T, T is Σ_1-sound, and g is primitive recursive.
  (a) Show that there are true Σ_1 sentences σ_0, σ_1 and a proof p of $Pr(\sigma_0) \lor Pr(\sigma_1)$ such that neither $Pr(\sigma_0)$ nor $Pr(\sigma_1)$ has a proof ⩽ g(p). [Hint:

Show that there is a primitive recursive function h such that h is provably increasing in T and if

(*) $\quad T \vdash \sigma \leftrightarrow \exists z (Prf(Pr(\chi),z) \wedge \forall u \leqslant z \neg Prf(Pr(\sigma),u)),$
$\quad T \vdash \chi \leftrightarrow \exists z (Prf(Pr(\sigma) \vee Pr(\chi),z) \wedge \forall u \leqslant h(g(z)) \neg Prf(\sigma,u)),$

r is a proof of $Pr(\chi)$, and $Pr(\sigma)$ has no proof $\leqslant r$, then there is a proof $\leqslant h(r)$ of σ. (Analyze the proof of Lemma 1.1(c).) Let $\sigma_0 := \sigma$ and $\sigma_1 := \chi$. Use Löb's theorem to show that $T \vdash Pr(\sigma_0) \vee Pr(\sigma_1)$.]

(b) Show that there are Σ_1 sentences χ_0, χ_1 such that χ_0 is true, χ_1 is false (in fact, $T \vdash \neg \chi_1$) and $Pr(\chi_0) \vee Pr(\chi_1)$ has a proof p such that $Pr(\chi_0)$ has no proof $\leqslant g(p)$. [Hint: In (*) replace $Pr(\chi)$ by $Pr(\sigma^*)$, where $\sigma^* :=$
$\exists u(Prf(Pr(\sigma),u) \wedge \forall z < u \neg Prf(Pr(\chi),z))$.
Let $\chi_0 := \sigma$ and $\chi_1 := \sigma^*$.]

36. Suppose PA ⊣ T and T is Σ_1-sound.

(a) Show that Theorem 14 and Exercises 34, 35 hold with "primitive recursive" replaced by "provably recursive in T".

(b) There is a recursive function f such that if p is a proof of the sentence $Pr(\varphi)$, then f(p) is a proof of φ. Show that f is not provably recursive in T.

Notes for Chapter 2

Theorem 1 is due to Gödel (1931). (However, Gödel assumed that T is ω-consistent (see Exercise 30) but then applied this assumption only to the formula (corresponding to) $Prf_T(\varphi,x)$.) The essential content of Gödel's theorem is that truth and provability in arithmetic are not equivalent (or: the set of true sentences of L_A is not r.e.). This also follows from each of Exercises 2–5 and 7. Theorem 2 is due to Rosser (1936). For alternative proofs of Theorem 1 and a result similar to Theorem 2, see Kotlarski (1994), (1996). Theorems 1 and 2 can be strengthened and generalized in a number of different directions as indicated in Exercises 1–3 (see also Chapter 8). However, these "directions" lead away from the central theme of this book and so will not be pursued further; but see, for example, Kleene (1952a), Mostowski (1952b), (1961), and Kreisel and Lévy (1968). Lemma 1 is due to Lindström (1979). Theorem 3 is due to Mostowski (1961); for a stronger result also due to Mostowski (1961), see Exercise 21.

Theorem 4 is essentially due to Gödel (1931); the present general formulation is due to Feferman (1960) (cf. also Kotlarski (1994), (1998)). An interesting account of the events and ideas behind the seminal paper

Feferman (1960) is given in Feferman (1997). Corollary 1 is due to Mostowski (1952a) and Ryll-Nardzewski (1952); this result is strengthened in Chapter 4 (Corollary 4.1) and Chapter 6 (Theorem 6.3). Theorem 6 is due to Löb (1955). Löb's theorem or, more exactly, (L2) is one of the keys to the logic of provability (cf. Boolos (1979), (1993), Smoryński (1985), Lindström (1996), Japaridze and de Jongh (1998)). Theorem 7 is due to Feferman (1960). Corollary 4 is due to Feferman (1960), (1997); for a stronger result, also due to Feferman (1960), see Exercise 6.3(b). Theorem 8(a) (with a different proof) is due to Feferman (1960); Theorem 8(b) is due to Orey (see Feferman (1960)).

Theorem 9 is due to Mostowski (1961) (cf. also Myhill (1972)). Theorem 10 is due to Scott (1962). Theorem 11 (with a different proof) is due to Montagna (1982).

A result similar to Theorem 13 was first obtained by Gödel (1936) (cf. also Mostowski (1952b) and Ehrenfeucht and Mycielski (1971)). Theorem 14, improved as in Exercise 36(a), is due to Parikh (1971); the present proof was pointed out to me by Christian Bennet; see also de Jongh and Montagna (1989); a more general result has been proved by Montagna (1992); cf. also Hájek, Montagna, Pudlák (1992); for related results, see Exercise 5.10.

The recursion-theoretic proofs of the incompleteness theorems of Gödel and Rosser, Exercises 2(a) and 3(b), are due to Kleene (cf. Kleene (1952a)). Exercise 11 is due to Montague (1963). Exercise 14(a) is due to Kreisel (see Smoryński (1985)). Exercise 15(a) and (b) are special cases of a general result, the fixed point theorem of provability logic (a subject not treated in this book) due to Dick de Jongh (unpublished) and Sambin (1976) (cf. also Boolos (1979), (1993), Smoryński (1985), Lindström (1996)). Exercise 17(b) is due to Feferman (1960); it was used by him to prove Theorem 6.8. Exercise 19 is due to Hájkova (1971); her papers contain many related results. Exercise 21 is due to Mostowski (1961). Exercise 24 (with a different proof) is due to Kripke (1963). The equivalence of (i), (ii), (iii) in Exercise 25 is due to Jensen and Ehrenfeucht (1976) and Guaspari (1979); for similar results, see Exercise 5.2. Exercise 26(a) is due to Harvey Friedman, Warren Goldfarb, and Leo Harrington (independently and unpublished; cf. Smoryński (1981b)); Exercise 26(b) is due to Volodya Shavrukov (unpublished). Exercise 27(a),(b),(c) are due to Bennet (1986); his book contains a number of further results on the relation ≼. Exercise 35 improved as in Exercise 36(a), is due to Shavrukov (1993) (with different proofs).

3. NUMERATIONS OF R.E. SETS

Any set numerated in T is r.e. The question arises if the converse of this is true, in other words, if every r.e. set can be numerated in T. If T is Σ_1-sound, then, of course, the answer is "yes" (Corollary 1.4). If T is not Σ_1-sound, the answer is still "yes" although this is not so obvious. This is the first and most important result of this chapter. We also prove some refinements of this result.

Beginning in this chapter we omit most references to the Lemmas, Facts, and Corollaries of Chapter 1. To avoid too much repetition, proofs are sometimes left to the reader.

§1. Numerations of r.e. sets. Let X be any r.e. set. Our first task is to show that X can be numerated in T even if T is not Σ_1-sound. We have already solved a similar problem in generalizing Gödel's incompleteness theorem to non Σ_1-sound theories (Theorem 2.2). A similar construction will suffice for our present problem.

Theorem 1. Let X be any r.e. set. There is then a Σ_1 (Π_1) formula $\xi(x)$ which numerates X in T.

Proof. There is a primitive recursive relation R(k,m) such that X = {k: $\exists$m R(k,m)}. Let $\rho(x,y)$ be a PR binumeration of R(k,m). Let $\xi(x)$ be such that
(1) $\quad Q \vdash \xi(k) \leftrightarrow \exists y(\rho(k,y) \wedge \forall z \leq y \neg \mathrm{Prf}_T(\xi(k),z))$.
Then $\xi(x)$ is Σ_1. We are going to show that $\xi(x)$ numerates X in T.
  Suppose first $k \in X$. There is then an m such that
(2) $\quad Q \vdash \rho(k,m)$.
Now, for *reductio ad absurdum*, suppose $T \nvdash \xi(k)$. Then $Q \vdash \neg \mathrm{Prf}_T(\xi(k),p)$ for every p. It follows that
(3) $\quad Q \vdash \forall z \leq m \neg \mathrm{Prf}_T(\xi(k),z)$.
Combining (2) and (3) we get
  $Q \vdash \exists y(\rho(k,y) \wedge \forall z \leq y \neg \mathrm{Prf}_T(\xi(k),z))$.
But then, by (1), $Q \vdash \xi(k)$ and so $T \vdash \xi(k)$ and we have reached the desired contradiction. Thus, $T \vdash \xi(k)$.
  Next suppose $T \vdash \xi(k)$. Let p be a proof of $\xi(k)$ in T. Then $Q \vdash \mathrm{Prf}_T(\xi(k),p)$ and so

(4) $Q \vdash \forall z \leqslant y \neg \mathrm{Prf}_T(\xi(k),z) \to y < p$.

Suppose $k \notin X$. Then $Q \vdash \neg \rho(k,m)$ for every m. It follows that

(5) $Q \vdash \neg \exists y < p \rho(k,y)$.

Combining (4) and (5) we get

$$Q \vdash \neg \exists y (\rho(k,y) \wedge \forall z \leqslant y \neg \mathrm{Prf}_T(\xi(k),z)),$$

whence, by (1), $Q \vdash \neg \xi(k)$ and so $T \vdash \neg \xi(k)$, impossible. Thus, $k \in X$ and we have shown that $\xi(x)$ numerates X in T.

Next let $\xi(x)$ be such that

$$Q \vdash \xi(k) \leftrightarrow \forall y (\mathrm{Prf}_T(\xi(k),y) \to \exists z \leqslant y \rho(k,z)).$$

Then $\xi(x)$ is Π_1. We leave the proof that $\xi(x)$ numerates X in T to the reader. ∎

Let us say that $\xi(x)$ *correctly numerates* X *in* T if $\xi(x)$ numerates X in T and for every k, $T \vdash \xi(k)$ iff $\xi(k)$ is true. We can partially improve Theorem 1 as follows.

Theorem 1'. *The Σ_1 formula $\xi(x)$ defined in the proof of Theorem 1 numerates X correctly in T.*

Proof. If $\xi(k)$ is true, then $T \vdash \xi(k)$, since $\xi(x)$ is Σ_1. Conversely, suppose $T \vdash \xi(k)$. Let p be the least proof of $\xi(k)$ in T. Then (4) holds. Suppose there is no m < p such that $R(k,m)$. Then (5) follows and so, as before, we get $T \vdash \neg \xi(k)$, which is impossible. Thus, there is an m < p such that $R(k,m)$. But then $\rho(k,m)$ is true. Also, p being minimal, $\forall z \leqslant m \neg \mathrm{Prf}_T(\xi(k),z)$ is true. It follows that

$$\exists y (\rho(k,y) \wedge \forall z \leqslant y \neg \mathrm{Prf}_T(\xi(k),z))$$

is true and so, by (1), $\xi(k)$ is true, as desired. ∎

Note that if X is numerated correctly in T by a Π_1 formula, then X is recursive.

The following strengthening of Theorem 1 is occasionally useful.

Lemma 1. *Suppose X and Y are r.e. and Y is monoconsistent with Q. There is then a Σ_1 (Π_1) formula $\xi(x)$ such that for every k, if $k \in X$, then $Q \vdash \xi(k)$ and if $k \notin X$, then $\xi(k) \notin Y$.*

The proof is again left to the reader. Lemma 1 also follows from Lemma 2(a), below.

We now ask if there are (Σ_1) formulas $\xi(x)$ which not only numerate X in T but also satisfy additional conditions in terms of provability or non-provability of (propositional combinations of) sentences of the form $\xi(k)$

with $k \notin X$. The following result is a first step in this direction.

Theorem 2. Let X_0 and X_1 be disjoint r.e. sets. There is then a Σ_1 formula $\xi(x)$ such that $\xi(x)$ numerates X_0 in T and $\neg\xi(x)$ numerates X_1 in T.

Proof. Let $R_i(k,m)$ be a primitive recursive relation such that $X_i = \{k: \exists m R_i(k,m)\}$, $i = 0, 1$. Let $\rho_i(x,y)$ be a PR binumeration of $R_i(k,m)$. Let $\xi(x)$ be such that
(1) $\quad Q \vdash \xi(k) \leftrightarrow \exists y((\rho_0(k,y) \vee \mathrm{Prf}_T(\neg\xi(k),y)) \wedge$
$$\forall z \leqslant y(\neg\rho_1(k,z) \wedge \neg\mathrm{Prf}_T(\xi(k),z))).$$
We show that $\neg\xi(x)$ numerates X_1 in T; the proof that $\xi(x)$ numerates X_0 in T is similar and is left to the reader.

Suppose first $k \in X_1$ and, for *reductio ad absurdum*, $T \nvdash \neg\xi(k)$. Then there is an m such that $T \vdash \rho_1(k,m)$ and $T \vdash \neg\mathrm{Prf}_T(\neg\xi(k),p)$ for all p. Also $k \notin X_0$ and so $T \vdash \neg\rho_0(k,n)$ for all n. It follows that
$$T \vdash \forall z \leqslant y \neg\rho_1(k,z) \to y < m,$$
$$T \vdash \neg\exists y < m (\rho_0(k,y) \vee \mathrm{Prf}_T(\neg\xi(k),y)).$$
But then, by (1), $T \vdash \neg\xi(k)$, contrary to assumption. Thus, $T \vdash \neg\xi(k)$.

Next suppose $T \vdash \neg\xi(k)$ and let p be such that $T \vdash \mathrm{Prf}_T(\neg\xi(k),p)$. We also have and $T \nvdash \xi(k)$ and so $T \vdash \neg\mathrm{Prf}_T(\xi(k),m)$ for all m. Suppose now $k \notin X_1$. Then $T \vdash \neg\rho_1(k,m)$ for all m. It follows that
$$T \vdash \mathrm{Prf}_T(\neg\xi(k),p)) \wedge \forall z \leqslant p(\neg\rho_1(k,y) \wedge \neg\mathrm{Prf}_T(\xi(k),z)).$$
But then, by (1), $T \vdash \xi(k)$, impossible. Thus, $k \in X_1$. ∎

One aspect of the above question is to ask to what extent results on numerations of r.e. sets can be combined with results on independent formulas. For example, does there exist a (Σ_1) formula $\xi(x)$ which numerates X in T and is *independent on* X^c ($= N - X$) *over* T in the sense that the only propositional combinations of sentences $\xi(k)$, with $k \in X^c$, provable in T are the tautologies? We now show that the answer is affirmative.

To prove this we need part (a) of the following lemma; Lemma 2(b) will be needed later, in the proof of Theorem 7.13.

Lemma 2. Suppose X and Y are r.e. and Y is monoconsistent with Q.

(a) There is then a Σ_1 (Π_1) formula $\xi(x)$ numerating X in Q and such that
(*) $\quad$ for every finite subset X' of X^c, $\vee \{\xi(k): k \in X'\} \notin Y$.

(b) Suppose PA ⊣ T. There are then formulas $\xi(x)$ and $\xi'(x)$ such that $\xi(x)$ is Π_1, $\xi'(x)$ is Σ_1, PA $\vdash \xi'(x) \to \xi(x)$, $\xi'(x)$ numerates X in Q, and (*) holds.

§1. *Numerations of R.E. Sets*   53

Proof. We may assume that $\mathrm{Th}(Q) \subseteq Y$. (If necessary, replace Y by $\mathrm{Th}(Q) \cup Y$; this set is still monoconsistent with Q.) Let $R(k,m)$ and $R^*(k,m)$ be primitive recursive relations such that $X = \{k: \exists m R(k,m)\}$ and $Y = \{k: \exists m R^*(k,m)\}$ and let $\rho(x,y)$ be a PR binumeration of $R(k,m)$. Let $S(\eta,m)$ be the following primitive recursive relation:

    there are $r \leq m$ and $k_0,\ldots,k_r \leq m$ such that $R^*(\bigvee\{\eta(k_s): s \leq r\},m)$
    and $\forall s \leq r \forall p \leq m \neg R(k_s,p)$.

Let $\sigma(x,y)$ be a PR binumeration of $S(\eta,m)$.

(a) We construct a Σ_1 formula as desired. Let $\xi(x)$ be such that

(1) $Q \vdash \xi(x) \leftrightarrow \exists z(\rho(x,z) \wedge \forall u \leq z \neg \sigma(\xi,u))$.

We now show that

(2) $\neg S(\xi,m)$ for every m.

Suppose $S(\xi,m)$ is true. Then $Q \vdash \sigma(\xi,m)$. Hence, by (1), for every k,

(3) $Q \vdash \xi(k) \to \exists z \leq m \rho(k,z)$.

Moreover, there are $r \leq m$ and $k_0,\ldots,k_r \leq m$ such that $\bigvee\{\xi(k_s): s \leq r\} \in Y$ and $\forall s \leq r \forall p \leq m \neg R(k_s,p)$. It follows that $\forall s \leq r \forall p \leq m Q \vdash \neg \rho(k_s,p)$. But then, by (3), $Q \vdash \neg \bigvee\{\xi(k_s): s \leq r\}$, contradicting the fact that Y is monoconsistent with Q. This proves (2).

We may assume that if $R^*(\bigvee\{\eta(k_s): s \leq r\},m)$, then $r \leq m$ and $k_s \leq m$ for $s \leq r$. But then (*) follows at once from (2). Finally, the fact that $\xi(x)$ numerates X in T follows from (1), (2), (*), since $\mathrm{Th}(Q) \subseteq Y$.

Next let $\xi(x)$ be such that

(4) $Q \vdash \xi(x) \leftrightarrow \forall u(\sigma(\xi,u) \to \exists z \leq u \rho(x,z))$.

Then $\xi(x)$ is Π_1 and has the desired properties; details are left to the reader. ♦

(b) Let $\xi(x)$ be the formula defined by (4) and let $\xi'(x) :=$
    $\exists z(\rho(x,z) \wedge \forall u \leq z \neg \sigma(\xi,u))$.
The verification that $\xi(x)$ and $\xi'(x)$ are as claimed should now be easy. ∎

Lemma 2(b) can also be obtained as an easy consequence of Theorem 5, below.

Theorem 3. Let X be any r.e. set. There is then a Π_1 (Σ_1) formula $\eta(x)$ which numerates X in T and is independent on X^c over T.

Proof. By Theorem 2.9, there is a Π_1 (Σ_1) formula $\mu(x)$ which is independent (on N) over T. Let
    $Y = \bigcup\{\mathrm{Th}(T + \{\mu^{f(k)}(k): k \leq n \,\&\, f \in 2^{n+1}\}): n \in N\}$.
Then Y is r.e. and monoconsistent with Q. Let $\xi(x)$ be the Π_1 (Σ_1) formula given by Lemma 2(a). Let $\eta(x) := \xi(x) \vee \mu(x)$.

If $k \in X$, then $Q \vdash \eta(k)$. To see that $\eta(x)$ is independent on X^c, suppose, for example, that $k_0, k_1, k_2, k_3 \in X^c$ are distinct and that
$$T \vdash \eta(k_0) \vee \eta(k_1) \vee \neg\eta(k_2) \vee \neg\eta(k_3).$$
Then
$$T + \neg\mu(k_0) + \neg\mu(k_1) + \mu(k_2) + \mu(k_3) \vdash \xi(k_0) \vee \xi(k_1),$$
contrary to the choice of $\xi(x)$. ∎

In Chapter 4 Theorem 3 will be used to construct not irredundantly axiomatizable theories (Theorem 4.7).

§2. Types of independence.

By a *type* (*of independence*) we understand a consistent r.e. set P of propositional formulas F in the propositional variables p_n, $n \in N$, closed under tautological consequence. Let $\langle \varphi_k : k < \omega \rangle$ be a sequence of sentences. Let $F(\langle \varphi_k : k < \omega \rangle)$ be obtained from F by replacing p_k by φ_k for each k. If $\xi(x)$ is a formula, let $F(\xi) = F(\langle \xi(k) : k < \omega \rangle)$. $\langle \varphi_k : k < \omega \rangle$ is of *type* P *over* T if
$$P = \{F : T \vdash F(\langle \varphi_k : k < \omega \rangle)\}.$$
$\xi(x)$ is of *type* P *over* T if this is true of $\langle \xi(k) : k < \omega \rangle$.

Theorem 4. For every type P, there is a primitive recursive sequence $\langle \varphi_k : k < \omega \rangle$ of B_1 sentences of type P over T.

Proof. In what follows p_k^i is p_k, if $i = 0$, and $\neg p_k$, if $i = 1$. Let s be a sequence of 0's and 1's, $s = \langle i_0, \ldots, i_k \rangle$. Then F^s is $p_0^{i_0} \wedge \ldots \wedge p_k^{i_k}$. Assuming that $\varphi_0, \ldots, \varphi_k$ have been defined, we define φ^s in a similar manner.

We now define $\varphi_0, \varphi_1, \varphi_2, \ldots$ It will be clear that the sentences φ_k are B_1 and that the sequence $\langle \varphi_k : k < \omega \rangle$ is primitive recursive. In addition to this it is sufficient to guarantee that for every k and every $s = \langle i_0, \ldots, i_k \rangle$,
(1) $P + F^s$ is consistent iff $T + \varphi^s$ is consistent.
Without loss of generality we may assume that $p_0 \in P$. Let $\varphi_0 := T$. Then (1) holds for $k = 0$. Suppose (1) holds for $k = n$. Let
$$Y_0^s = \{m : (F^s \to p_m) \in P\}, \quad Y_1^s = \{m : (F^s \to \neg p_m) \in P\}.$$
Next let $\xi_s(x)$ be a Σ_1 formula defined as in the proof of Theorem 2 with Y_0 and Y_1 replaced by Y_0^s and Y_1^s and T replaced by $T + \varphi^s$. Then
(2)   if $P + F^s$ is consistent, then Y_0^s and Y_1^s are disjoint,
(3)   if $T + \varphi^s$ is consistent and Y_0^s and Y_1^s are disjoint, then
$$Y_0^s = \{m : T + \varphi^s \vdash \xi_s(m)\}, \quad Y_1^s = \{m : T + \varphi^s \vdash \neg\xi_s(m)\}.$$
Let $s_0, \ldots, s_q$ be all sequences of 0's and 1's of length $n+1$. Finally, set
$$\varphi_{n+1} := (\varphi^{s_0} \wedge \xi_{s_0}(n+1)) \vee \ldots \vee (\varphi^{s_q} \wedge \xi_{s_q}(n+1)).$$

Then

(4) $\quad T + \varphi^s \vdash \varphi_{n+1} \leftrightarrow \xi_s(n+1)$.

To complete the induction, we now have to show that

(5) $\quad P + F^s \wedge p_{n+1}$ is consistent iff $T + \varphi^s \wedge \varphi_{n+1}$ is consistent,

(6) $\quad P + F^s \wedge \neg p_{n+1}$ is consistent iff $T + \varphi^s \wedge \neg\varphi_{n+1}$ is consistent.

To prove (5), suppose first $P + F^s \wedge p_{n+1}$ is consistent. Then $n+1 \notin Y_1^s$. Moreover, $P + F^s$ is consistent and so, by (2), Y_0^s and Y_1^s are disjoint and, by the inductive assumption, $T + \varphi^s$ is consistent. It follows, by (3), that $T + \varphi^s \nvdash \neg\xi_s(n+1)$ and so, by (4), $T + \varphi^s \wedge \varphi_{n+1}$ is consistent.

Next suppose $T + \varphi^s \wedge \varphi_{n+1}$ is consistent. Then $P + F^s$ is consistent. Hence, by (2), (3), (4), $n+1 \notin Y_1^s$ and so $P + F^s \wedge p_{n+1}$ is consistent.

This proves (5). The proof of (6) is similar. ∎

From Theorem 4 and Fact 10(b) we get:

Corollary 1. Suppose PA ⊣ T. Then for each type P, there is a Δ_2 formula of type P over T

Suppose T is Σ_1-sound, $\xi(x)$ is Σ_1, and $\xi(x)$ is of type P over T. Then P is *positively prime* (*p.p.*) in the sense that for all propositional variables $p_{n_0},...,p_{n_k}$, if $p_{n_0} \vee ... \vee p_{n_k} \in P$, there is an $i \leq k$ such that $p_{n_i} \in P$. (A formula F is *p.p.* if the set of tautological consequences of F is p.p.) We now prove that, for extensions of PA, the converse of this is true.

Theorem 5. Suppose PA ⊣ T. Then for each p.p. type P, there is a Σ_1 formula of type P over T.

In the proof of Theorem 5 we shall have to rely on the reader's ability to formalize (fairly simple) intuitive arguments and definitions in PA (or willingness to believe that these arguments and definitions can be so formalized). It will be essential to distinguish between the claims (i): for every k, PA proves: ...k... and (ii): PA proves: for every k, ...k... Here (ii) is the stronger claim; it may very well be the case that (i) is true and (ii) is false.

We are going to define a certain primitive recursive function f(k,m,n); the details of the definition will be crucial. The (inductive) definition of the function f(k,m,n) is given in the metalanguage and the task of formalizing this definition is left to the reader.

The numbers 0, 1 will be thought of as the truth-values *falsity* and *truth*, respectively. A function $t \in 2^N$ can then be regarded as a truth-value assignment: t assigns truth to p_i iff $t(i) = 1$. We always assume that $t(i) = 0$ for all but finitely many i. Thus, t is essentially a finite object and can be coded by, and treated as, a natural number. t[F] is the truth-value assigned by t to the formula F; for example, $t[p_i] = t(i)$.

By induction on the length of F, it is easy to show that for every F,

(1) PA proves: if for every i such that p_i occurs in F, $\xi(i)$ iff $t(i) = 1$, then $F(\xi)$ iff $t[F] = 1$.

Suppose g, $h \in 2^N$. Then g precedes h in the *lexiographic ordering* if $g \neq h$ and $g(k) < h(k)$, where k is the least number such that $g(k) \neq h(k)$. We shall also use the following partial ordering of 2^N: $g \leq h$ iff $g(k) \leq h(k)$ for all k.

Lemma 3. (a) Suppose P is p.p. Then there is a primitive recursive function $G(s)$ such that (i) P is tautologically equivalent to $\{G(s): s \in N\}$, (ii) for every s, $G(s)$ is p.p., and (iii) PA proves: for all s, s', if $s < s'$, then $G(s') \to G(s)$ is a tautology (we may assume that $G(0)$ is a tautology), (iv) PA proves: for all i, s, if p_i occurs in $G(s)$, then $i \leq s$.

(b) If F is p.p. and consistent, there is a $\leq$-least t such that $t[F] = 1$.

(c) Let t^s be the $\leq$-least t such that $t[G(s)] = 1$. For every s, $t^s \leq t^{s+1}$.

Proof. (a) Let $F_0, F_1, F_2, \ldots$ be a primitive recursive enumeration of P (possibly with repetitions) such that (provably in PA) if p_i occurs in F_s, then $i \leq s$. Let $Y(s) = \{F_j: j \leq s\}$. Then $Y(s)$ is primitive recursive. (Finite sets of formulas can be coded by natural numbers.) We now define $Z(s)$ and an auxiliary function h as follows:

Stage 0. $Z(0) = \emptyset$, $h(0) = 0$.

Stage s+1. Case 1. There is a (least) p.p. set U such that $Z(s) \cup \{F_{h(s)}\} \subseteq U \subseteq Y(s)$. Let $Z(s+1) = U$, $h(s+1) = h(s)+1$.

Case 2. Otherwise. Let $Z(s+1) = Z(s)$, $h(s+1) = h(s)$.

Then $Z(s) \subseteq Z(s+1)$ for all s and $P = \bigcup\{Z(s): s \in N\}$. Finally, let $G(s) = \bigwedge Z(s)$. The verification that $G(s)$ is as claimed is left to the reader. ♦

(b) Suppose p_i occurs in F only if $i \leq n$. We can then ignore the p_i for which $i > n$. If $F \to p_0 \wedge \ldots \wedge p_n$ is a tautology, let $t(i) = 1$ for $i \leq n$. Otherwise let $I = \{i \leq n: F \to p_i$ is not a tautology$\}$. Suppose t is such that $t[F \wedge \bigwedge\{\neg p_i: i \in I\}] = 1$. Then t is as desired. If there is no t with this property, then $F \to \bigvee\{p_i: i \in I\}$ is a tautology. But then $F \to p_j$ is a tautology for some $j \in I$, a contradiction. ♦

(c) This is clear, since, by (a) (iii), $t^{s+1}[G(s)] = 1$. ∎

Proof of Theorem 5. Let $f(s,m,i)$ be the primitive recursive function defined below; m will always be assumed to be a formula $\eta(x)$, the value of

§2. *Types of Independence* 57

f(s,m,i), when m is not a formula, is irrelevant and we may set f(s,m,i) = 0.
Now let $\xi(x)$ be such that
(2) $PA \vdash \xi(x) \leftrightarrow \exists z(f(z,\xi,x) = 1)$
and let $h(s,i) = f(s,\xi,i)$. Also, let h_s be such that for all i,
   $h_s(i) = h(s,i)$.
$h_s(i)$ may be thought of as the truth-value assigned to p_i at stage s. It will be clear from the definition of f that for fixed η and i, $f(s,\eta,i)$ is nondecreasing in s. Thus, informally, $\xi(i)$ is true iff the truth-value eventually assigned to p_i is 1.

Our goal is to define f in such a way that the following two claims can be established; G(s) is as in Lemma 3(a).

Claim 1. For every s, $PA \vdash (G(s))(\xi)$.

Claim 2. For every F, if $T \vdash F(\xi)$, then $F \in P$.

By Lemma 3(a)(i), Theorem 5 follows from Claims 1 and 2.

Cases 1.1 and 2 of the definition of f are designed to ensure the validity of Claim 2: If $T \vdash F(\xi)$ and, for a suitable s, $G(s) \to F$ is not a tautology, Case 1.1 applies at Stage s+1 and so $h_{s+1}[F] = 0$. Also Case 2 applies at all later stages and so $h_{s'} = h_{s+1}$, whence $h_{s'}[F] = 0$, for all $s' > s$. This is provable in PA. It follows, by (1), that $PA \vdash \neg F(\xi)$, contradicting the assumption that $T \vdash F(\xi)$.

We now define $f(s,\eta,i)$ and at the same time a (trivial) auxiliary function $g(s,\eta)$ as follows:

Stage 0. $f(0,\eta,i) = g(0,\eta) = 0$.

Stage s+1. Case 1. $g(s,\eta) = 0$.

  Case 1.1. $s = \langle F,m \rangle$, m is a proof of $F(\eta)$ in T, and there is a t such that
(3) $t[G(s)] = 1$,
(4) $t[F] = 0$,
(5) $f(s,\eta,i) \leq t(i)$ for $i < s$.
Let t' be the lexicographically least such t. $g(s+1,\eta) = 1$ and
   $f(s+1,\eta,i) = t'(i)$.

  Case 1.2. Not Case 1.1 and there is a t such that (5) holds and
(6) $t[G(s+1)] = 1$.
Let t' be the lexicographically least such t. $g(s+1,\eta) = 0$ and
   $f(s+1,\eta,i) = t'(i)$.

  Case 1.3. Otherwise. $g(s+1,\eta) = 0$ and
   $f(s+1,\eta,i) = f(s,\eta,i)$.

Case 2. $g(s,\eta) \neq 0$. $g(s+1,\eta) = 1$ and
   $f(s+1,\eta,i) = f(s,\eta,i)$.

Inspection of the above definition in conjunction with Lemma 3(a)(iv) shows that $h_s(i) = 0$ whenever $i > s$; in fact, this can easily be proved in PA, in other words:

(7) PA proves: for all i and s, if $i > s$, then $h_s(i) = 0$.

Furthermore,

(8) if $s < s'$, then $h_s \preceq h_{s'}$.

For $s' = s+1$, this can be seen by inspection; the full result follows by induction.

Using (7), the proof of (8) can be formalized in PA and so we have

(9) PA proves: for all s, s', if $s < s'$, then $h_s \preceq h_{s'}$.

Next we show that

(10) for every s, $g(s,\xi) = 0$; in other words, if $\eta := \xi$, Case 1.1 never applies.

Suppose not and let s' be the least number such that $g(s',\xi) = 1$. Then Case 1.1 applies at Stage s'. Thus, $s'-1 = \langle F,m \rangle$, m is a proof of $F(\xi)$ in T, whence

(11) $T \vdash F(\xi)$,

and $h_{s'}[F] = 0$. Let $t' = h_{s'}$. For $\eta := \xi$ Case 2 now applies at every $s > s'$ and so $h_s = t'$ for every $s \geq s'$. By (8), $h_s \preceq t'$ for $s < s'$. It follows that $t'(i) = 1$ iff there is an s such that $h(s,i) = 1$.

Using (9), this argument can be formalized in PA and so

$$PA \vdash \exists z(h(z,x) = 1) \leftrightarrow t'(x) = 1.$$

But then, by (2), $PA \vdash \xi(x) \leftrightarrow t'(x) = 1$ and so, by (1), $PA \vdash F(\xi) \leftrightarrow t'[F] = 1$. But $PA \vdash t'[F] = 0$. It follows that $PA \vdash \neg F(\xi)$, contradicting (11). This proves (10).

We now show that for all s,

(12) $h_s = t^s$,

where t^s is as in Lemma 3(c). Since $G(0)$ is a tautology, this holds for $s = 0$. Suppose (12) holds for s. Then, by Lemma 3(c), $h_s \preceq t^{s+1}$. Since $t^{s+1}[G(s+1)] = 1$, either Case 1.1, Case 1.2 or Case 2 applies at $s+1$. By (8), Cases 1.1 and 2 don't and so Case 1.2 does. Also, the lexicographically least t' mentioned in Case 1.2 with $\eta := \xi$ is t^{s+1}. It follows that $h_{s+1} = t^{s+1}$. This proves (12).

From (7) and (12) it follows that

(13) for every s, PA proves: $h_s = t^s$.

Next we show that

(14) for every s, PA proves: for every $s' \geq s$, $h_{s'}[G(s)] = 1$.

Argue in PA: "For $s' = s$ we have $h_{s'} = t^s$, by (13), and so $h_{s'}[G(s)] = 1$. Suppose $s' \geq s$ and the statement holds for s'. If Case 1.3 or Case 2 applies at $s'+1$, then $h_{s'+1} = h_{s'}$ and so, by the inductive assumption, $h_{s'+1}[G(s)] = 1$. If Case 1.1 or Case 1.2 applies at $s'+1$, then $h_{s'+1}[G(s')] = 1$ or

$h_{s'+1}[G(s'+1)] = 1$ and so, by Lemma 3(a) (iii), $h_{s'+1}[G(s)] = 1$. Now the desired conclusion follows by induction." (Since this argument takes place in PA, Cases 1.1 and 2 cannot be ruled out.) This proves (14).

Proof of Claim 1. Fix s. Argue in PA: "By (2) and (9), there is an $s' \geq s$ such that for every $i \leq s$, $h_{s'}(i) = 1$ iff $\xi(i)$. By (14), $h_{s'}[G(s)] = 1$. By Lemma 3(a) (iv), no p_i with $i > s$ occurs in $G(s)$. Thus, by (1), $(G(s))(\xi)$."

Proof of Claim 2. Let m be a proof of $F(\xi)$ in T. Let $s = \langle F, m \rangle$. By Lemma 3(a) (i), it is sufficient to show that $G(s) \to F$ is a tautology. Suppose not. Let t be such that $t[G(s)] = 1$ and $t[F] = 0$. Then $t^s \leq t$. By (13), $h_s = t^s$ and so $h_s \leq t$. But then Case 1.1 applies at $s+1$ and so $g(s+1,\xi) = 1$, contrary to (10). Thus, $G(s) \to F$ is a tautology. Finally, $G(s) \in P$ and so $F \in P$.

This concludes the proof of Theorem 5. ∎

For PA ⊣ T, Theorems 1, 2, 3 are, of course, special cases of Theorem 5.

Exercises for Chapter 3

1. Suppose $Q \dashv T_0 \dashv T_1$. Show that for every r.e. set, there is a Σ_1 formula which numerates X in both T_0 and T_1.

2. We write $S \dashv_p T$ to mean that S is a proper subtheory of T.

    (a) Suppose $Q \dashv T_0 \dashv_p T_1$. Let X_0 and X_1 be r.e. sets such that $X_0 \subseteq X_1$. Show that there is a formula $\xi(x)$ numerating X_i in T_i, $i = 0, 1$. [Hint: Let θ be such that $T_0 \nvdash \theta$ and $T_1 \vdash \theta$. There exist a formula $\xi_1(x)$ numerating X_1 in T_0 and in T_1 and a formula $\xi_0(x)$ numerating X_0 in $T_0 + \neg \theta$. Let $\xi(x) := \xi_1(x) \wedge (\theta \vee \xi_0(x))$.]

    (b) Suppose $Q \dashv T_0 \dashv_p \ldots \dashv_p T_n$. Let X_i, $i \leq n$, be r.e. sets such that $X_i \subseteq X_{i+1}$ for $i < n$. Show that there is a formula $\xi(x)$ numerating X_i in T_i for $i \leq n$.

    (c) Suppose $Q \dashv T_0 \dashv T_1$ and suppose there is a formula $\sigma(x)$ which numerates Th(S) in S for every S such that $T_0 \dashv S \dashv T_1$. Show that $T_1 \dashv T_0$. [Hint: Suppose $T_1 \vdash \theta$ and let φ be such that $Q \vdash \varphi \leftrightarrow \neg\sigma(\varphi \vee \theta)$. Show that $T_0 \vdash \neg\varphi$.]

    (d) Suppose $Q \dashv T_0$, $Q \dashv T_1$, and T_0 and T_1 are incomparable (with respect to ⊣). Let X_0 and X_1 be any two r.e. sets. Show that there is a formula $\xi(x)$ which numerates X_i in T_i, $i = 0, 1$.

3. Suppose $Q \dashv T_0 \dashv_p T_1$. Show that there is a formula $\xi(x)$ such that for every recursive function f, the set

$\{n: T_0 \vdash \xi(n)$ & there is a proof p of $\xi(n)$ in T_1 such that $\xi(n)$ has no proof $\leq f(p)$ in $T_0\}$
is infinite, in fact, nonrecursive (this improves Theorem 2.13). [Hint: Let X be an r.e. nonrecursive set and let $\xi(x)$ be a formula numerating X in T_0 and N in T_1.]

4. Let X_0 and X_1 be r.e. sets. Let $\xi_0(x)$ be a Σ_n formula numerating X_0 in T. Show that there is a Σ_n formula $\xi_1(x)$ numerating X_1 in T such that $\xi_0(x) \vee \xi_1(x)$ numerates $X_0 \cup X_1$ in T. (If n = 1 and T is Σ_1-sound, this is trivial.) [Hint: Let $\rho(x,y)$ be a PR formula such that $\exists y \rho(x,y)$ correctly numerates X_1 in T, let $\xi(x)$ be such that
$$Q \vdash \xi(k) \leftrightarrow \exists y (\rho(k,y) \wedge \forall z \leq y \neg \text{Prf}_T(\xi(k) \vee \xi_0(k), z)),$$
and let $\xi_1(x) := \xi(x) \vee (\exists y \rho(x,y) \wedge \xi_0(x))$.]

5. Suppose PA ⊣ T. Let X be any r.e. set. Show that there is a Γ formula $\xi(x)$ numerating X in T and such that for every Γ formula $\eta(x)$, the theory $T + \{\xi(k) \leftrightarrow \eta(k): k \notin X\}$ is consistent. (For extensions of PA this improves Theorem 3.)

6. (a) Suppose PA ⊣ T and T is not Σ_1-sound. Show that the sentences φ_k in Theorem 4 can be taken to be Δ_1^T. [Hint: Use Lemma 1.3(vi).]
   (b) Suppose Q ⊣ S. Show that there are primitive recursive enumerations $\varphi_0, \varphi_1, \varphi_2, \ldots$ and $\psi_0, \psi_1, \psi_2, \ldots$ of *all* sentences such that the type of $\langle \varphi_n: n < \omega \rangle$ over S is the same as the type of $\langle \psi_n: n < \omega \rangle$ over T.

7. (a) Let $\rho_i(y)$, i = 0, 1, be PR formulas. Let φ be such that
$$Q \vdash \varphi \leftrightarrow \exists y ((\text{Prf}_T(\neg \varphi, y) \vee \rho_0(y)) \wedge \forall z \leq y (\neg \text{Prf}_T(\varphi, z) \wedge \neg \rho_1(z))).$$
Show that
   $T \vdash \varphi$ iff $\exists y (\rho_0(y) \wedge \forall z \leq y \neg \rho_1(z))$ is true,
   $T \vdash \neg \varphi$ iff $\exists z (\rho_1(z) \wedge \forall y < z \neg \rho_0(y))$ is true.
   (b) Derive Rosser's theorem (Theorem 2.2), Theorem 2, and Exercises 2.21, 2.22, 5.2(a) from (a).

Notes for Chapter 3

Theorems 1 and 2 are essentially due to Ehrenfeucht and Feferman (1960) and Putnam and Smullyan (1960), respectively; the present proofs are due to Shepherdson (1960). Lemmas 1 and 2 are due to Lindström (1979), (1984a).

Theorem 4 follows from a result of Pour-El and Kripke (1967) restricted to theories in L_A (see Exercise 6(b)); the proof is just an "effective" version of the proof that every denumerable Boolean algebra is embeddable in every (denumerable) atomless Boolean algebra. Theorem 5 is new; the proof is an adaption of a proof of Solovay (1985); the result solves Problem 32 of Friedman (1975); an interesting special case of Theorem 5 is proved in Montagna and Sorbi (1985).

Exercise 3 is due to di Paola (1975). Exercise 6(b) is a result of Pour-El and Kripke (1967) restricted to theories in L_A. Exercise 7(a) is the so called Shepherdson–Smoryński fixed point theorem (see Smoryński (1980) and Hájek and Pudlák (1993)); a more general result is proved in Smoryński (1981a).

4. AXIOMATIZATIONS

S is an *axiomatization* of T if S ⊣⊢ T. Suppose S ⊣ T. S + X is an *axiomatization of* T *over* S if X is r.e. and T ⊣⊢ S + X. In this chapter we discuss some important properties of axiomatizations: finiteness, boundedness, and irredundance.

§1. Finite and bounded axiomatizability; reflection principles. We shall say that T is a *finite extension of* S if there is a sentence φ such that T ⊣⊢ S + φ. T is *essentially infinite* (*e.i.*) *over* S if no consistent extension of T is finite over S. T is *e.i.* if T is e.i. over the empty theory (logic). We already know that PA is e.i. (Corollary 2.1).

By the *local reflection principle for* S we understand the set
$$\mathrm{Rfn}_S = \{\mathrm{Pr}_S(\varphi) \to \varphi : \varphi \text{ any sentence of } L_A\}.$$
Thus, Rfn_S is a piecemeal (local) way of saying that every sentence provable in S is true. (The latter statement, the full (global) reflection principle for S, cannot be expressed in T, since, by the Gödel–Tarski theorem, truth is not definable.)

Clearly PA + $\mathrm{Rfn}_T \vdash \mathrm{Con}_T$ (let $\varphi := \bot$). Also note that T is essentially reflexive iff T ⊢ $\mathrm{Rfn}_{T|k}$ for every k (cf. Corollary 1.8(b)).

We now use the local reflection principle to construct an e.i. extension of a given theory S. Note that T ⊢ Rfn_S implies T ⊢ S.

Theorem 1. If T ⊢ Rfn_S, then T is e.i. over S.

Proof. Suppose T ⊣ S + φ. We are going to show that S + φ is inconsistent. Let ψ be such that
(1) $Q \vdash \psi \leftrightarrow \neg\mathrm{Pr}_{S+\varphi}(\psi)$.
By hypothesis,
$$T \vdash \mathrm{Pr}_S(\varphi \to \psi) \to (\varphi \to \psi).$$
From this and (1) it follows that T ⊢ $\varphi \to \psi$. But then
(2)   S + $\varphi \vdash \psi$.
It follows that $Q \vdash \mathrm{Pr}_{S+\varphi}(\psi)$ and so, by (1), $Q \vdash \neg\psi$. But Q ⊣ S + φ and so, by (2), S + φ is inconsistent. ∎

If PA ⊣ T, the conclusion of Theorem 1 can be strengthened; see Corollary 2, below.

There is a stronger principle, the *uniform reflection principle*, which is a better approximation than Rfn_S of the full reflection principle for S:

$RFN_S = \{\forall x(\Gamma(x) \land Pr_S(x) \to Tr_\Gamma(x)): \Gamma \text{ arbitrary}\}$.

Clearly $T + RFN_S \vdash Rfn_S$ provided that $PA \dashv T$. Applying the uniform reflection principle we can derive a stronger conclusion than in Theorem 1.

A set X of sentences is *bounded* if $X \subseteq \Gamma$ for some Γ. T is *essentially unbounded* (*e.u.*) *over* S if for every bounded set X (not necessarily r.e), if $T \dashv S + X$, then $S + X$ is inconsistent. T is *e.u.* if T is e.u. over the empty theory.

Let $Prf_{S,\Gamma}(x,y) :=$
$\exists z(\Gamma(z) \land Tr_\Gamma(z) \land Prf_{S+z}(x,y))$
and let $Pr_{S,\Gamma}(x) := \exists y Prf_{S,\Gamma}(x,y)$.

Lemma 1. If φ is Γ, then
$$PA + RFN_S \vdash Pr_{S,\Gamma^d}(\varphi) \to \varphi.$$

Proof. Argue in $PA + RFN_S$: "Suppose $Pr_{S,\Gamma^d}(\varphi)$. There is then a Γ^d sentence ψ such that $Tr_{\Gamma^d}(\psi)$ and $Pr_S(\psi \to \varphi)$. By RFN_S,
$$\forall z(\Gamma(z) \land Pr_S(z) \to Tr_\Gamma(z)).$$
Since $\psi \to \varphi$ is Γ, it follows that $Tr_\Gamma(\psi \to \varphi)$. But $Tr_{\Gamma^d}(\psi)$. Consequently, by Fact 10(a) $Tr_\Gamma(\varphi)$ and so φ, as desired." ∎

Theorem 2. Suppose $PA \dashv T$ and $T \vdash RFN_S$. Then T is e.u. over S.

Proof. Suppose $X \subseteq \Sigma_n$ and $T \dashv S + X$. We are going to show that $S + X$ is inconsistent. Let ψ be such that
(1) $PA \vdash \psi \leftrightarrow \neg Pr_{S,\Sigma_n}(\psi)$.
Then ψ is Π_n. Hence, by Lemma 1,
$$T \vdash Pr_{S,\Sigma_n}(\psi) \to \psi.$$
From this and (1) it follows that $T \vdash \psi$ and so
(2) $S + X \vdash \psi$.
But then there is a conjunction θ of members of X such that $S + \theta \vdash \psi$. It follows that $T + \theta \vdash Tr_{\Sigma_n}(\theta) \land Pr_{S+\theta}(\psi)$ and so $T + \theta \vdash Pr_{S,\Sigma_n}(\psi)$, whence, by (1), $T + \theta \vdash \neg\psi$ and so $S + X \vdash \neg\psi$. Thus, by (2), $S + X$ is inconsistent. ∎

Note the obvious analogy between the proofs of Theorems 1 and 2, on the one hand, and the proof of Gödel's theorem (Theorem 2.1), on the other. Note also that if T is Σ_1-sound, then $X = \{\neg Pr_T(\varphi): T \nvdash \varphi\}$ is a (non-r.e.) set of Π_1 sentences such that $T + Rfn_T \dashv T + X$ and $T + X$ is consistent.

Since $PA \vdash RFN_\emptyset$ (Fact 11), we have (a) of the following corollary, improving Corollary 2.1.

Corollary 1. (a) PA is e.u.
 (b) If PA ⊣ T, then T + RFN_T is e.u. over T.

If PA ⊣ S, the above proof of Theorem 2 can be replaced by the following simple argument; the proof of Theorem 1 can be simplified in a similar way. Let
$$\chi := \forall x(\Gamma(x) \land \text{Pr}_S(x) \to \text{Tr}_\Gamma(x)).$$
Now let θ be any Γ^d sentence such that $S + \theta \vdash \chi$. Then $S + \theta \vdash \neg\text{Pr}_S(\neg\theta)$, whence $S + \theta \vdash \text{Con}_{S+\theta}$ and so $S + \theta$ is inconsistent, by Theorem 2.4.

This argument and (a somewhat more detailed version of) the above proof of Theorem 2 can be looked at from a different point of view which will be further elaborated in Chapter 5: Let φ be any Γ sentence such that $S + \neg\chi \vdash \varphi$. Then $S + \neg\varphi \vdash \chi$ and so $S \vdash \varphi$. Thus, $\neg\chi$ is Γ-conservative over S in the sense that if φ is any Γ sentence and $S + \neg\chi \vdash \varphi$, then $S \vdash \varphi$.

Next we show that if PA ⊣ T, no bounded r.e. extension of T is e.i. over T (and a bit more).

Theorem 3. Suppose PA ⊣ T, let X be an r.e. set of Γ sentences, and let Y be any r.e. set of sentences such that $T + X \nvdash \psi$ for every $\psi \in Y$. There is then a Γ sentence θ such that $T + \theta \vdash X$ and $T + \theta \nvdash \psi$ for every $\psi \in Y$.

Proof. By Craig's theorem, we may assume that X and Y are primitive recursive. Let $\xi(x)$ and $\eta(x)$ be PR binumerations of X and Y, respectively.
 Case 1. $\Gamma = \Pi_n$. Let θ be such that
$$\text{PA} \vdash \theta \leftrightarrow \forall y(\xi(y) \land \forall zu \leq y(\eta(z) \to \neg\text{Prf}_{T+\theta}(z,u)) \to \text{Tr}_{\Pi_n}(y)).$$
Suppose $\psi \in Y$ and $T + \theta \vdash \psi$. Let p be a proof of ψ in $T + \theta$; $\psi < p$. Then
$$\text{PA} \vdash \forall zu \leq y(\eta(z) \to \neg\text{Prf}_{T+\theta}(z,u)) \to y < p.$$
By Fact 10(a) (ii), it follows that $T + X|p \vdash \theta$ and so $T + X \vdash \psi$, contrary to hypothesis. Thus, $T + \theta \nvdash \psi$ for all $\psi \in Y$. But then
$$\text{PA} \vdash \forall zu \leq q(\eta(z) \to \neg\text{Prf}_{T+\theta}(z,u))$$
for all q. It follows that $T + \theta \vdash X$, as desired.
 Case 2. $\Gamma = \Sigma_n$. Let θ be such that
$$\text{PA} \vdash \theta \leftrightarrow \exists y(\exists zu \leq y(\eta(z) \land \text{Prf}_{T+\theta}(z,u)) \land \forall z \leq y(\xi(z) \to \text{Tr}_{\Sigma_n}(z))).$$
The verification that θ is as desired is left to the reader. ∎

From Theorem 1 and Theorem 3 with $Y = \{\bot\}$ we get the following:

Corollary 2. Suppose PA ⊣ T. If X is any bounded r.e. set of sentences such that $\text{Rfn}_T \dashv T + X$, then $T + X$ is inconsistent.

§1. *Finite and Bounded Axiomatizability; Reflection Principles* 65

Suppose T is Σ_1-sound. We have already mentioned that PA + $\text{Rfn}_T \vdash \text{Con}_T$. By Theorem 1, T + $\text{Con}_T \not\vdash \text{Rfn}_T$. Clearly PA + $\text{RFN}_T \vdash \text{Rfn}_T$. It has been pointed out that T + $\{\neg \text{Pr}_T(\varphi) : T \not\vdash \varphi\}$ is a consistent, bounded extension of T + Rfn_T. Thus, by Theorem 2, if PA ⊣ T, then T + $\text{Rfn}_T \not\vdash \text{RFN}_T$. These observations can be strengthened as follows.

We define the sentences Con(n,S), for n > 0, by: Con(1,S) := Con_S, Con(n+1,S) := Con(1,S + Con(n,S)). Let
$$\text{Con}_S^\omega = \{\text{Con}(n+1,S) : n \in \mathbb{N}\}.$$
The sets Rfn(n,S) are defined as follows: Rfn(0,S) = ∅, Rfn(1,S) := Rfn_S, Rfn(n+1,S) := Rfn(1,S + Rfn(n,S)). Next let
$$\text{Rfn}_S^\omega = \bigcup\{\text{Rfn}(n,S) : n \in \mathbb{N}\}.$$
We write S ⊣$_p$ S' to mean that S is a proper subtheory of S'.

Theorem 4. Suppose PA ⊣ T and T is Σ_1-sound.
(a) T + Con_T^ω ⊣$_p$ T + Rfn_T.
(b) T + Rfn_T^ω ⊣$_p$ T + RFN_T.

Lemma 2. (a) PA + $\text{Rfn}_T \vdash \text{Rfn}_{T+\text{Con}_T}$.
(b) PA + $\text{RFN}_T \vdash \text{RFN}_{T+\text{Rfn}_T}$.

Proof. (a) Let φ be any sentence.
$$\text{PA} + \text{Rfn}_T \vdash \text{Pr}_T(\text{Con}_T \to \varphi) \to (\text{Con}_T \to \varphi).$$
But, as we have already observed, PA + $\text{Rfn}_T \vdash \text{Con}_T$. It follows that PA + $\text{Rfn}_T \vdash \text{Pr}_{T+\text{Con}_T}(\varphi) \to \varphi$, as desired. ♦

(b) We give an informal proof using the fact that Fact 10(a) is provable in PA. We assume, as we may, that the PR binumeration $\rho(x)$ of Rfn_T implicit in the notation $\text{RFN}_{T+\text{Rfn}_T}$ is such that PA proves that every sentence satisfying $\rho(x)$ is of the form $\text{Pr}_T(\theta) \to \theta$. Suppose $\Sigma_1 \subseteq \Gamma$. Now argue in PA + RFN_T: "Let ψ be any Γ sentence provable in T + Rfn_T and let $\text{Pr}_T(\varphi_i) \to \varphi_i$, for i ⩽ n, be the members of Rfn_T occurring in the proof. We may assume that $\neg\text{Pr}_T(\varphi_i)$, for i ⩽ n, since those $\text{Pr}_T(\varphi) \to \varphi$ for which $\text{Pr}_T(\varphi)$ are provable in T and we may add the proofs of them to the original proof. Since $\neg\text{Pr}_T(\varphi_i) \to (\text{Pr}_T(\varphi_i) \to \varphi_i)$ is (trivially) provable in T, it follows that $\theta :=$
$$\neg\text{Pr}_T(\varphi_0) \wedge \ldots \wedge \neg\text{Pr}_T(\varphi_n) \to \psi$$
is provable in T. By RFN_T, $\text{Tr}_\Gamma(\theta)$. But, by Fact 10(a) (ii), $\text{Tr}_{\Gamma^d}(\neg\text{Pr}_T(\varphi_i))$, for i ⩽ n. Hence, by Fact 10(a) (iii), $\text{Tr}_\Gamma(\psi)$, as desired." ∎

Proof of Theorem 4. (a) In view of Lemma 2(a), it follows, by induction, that T + $\text{Rfn}_T \vdash \text{Con}_T^\omega$. T + Con_T^ω is consistent, since T is Σ_1-sound, and Con_T^ω

is an r.e. set of Π_1 sentences. Thus, by Corollary 2, $T + \text{Con}_T^\omega \nvdash \text{Rfn}_T$. ♦

(b) By Lemma 2(b), $T + \text{RFN}_T \vdash \text{Rfn}_T^\omega$. Let
$$X_k = \{\neg \text{Pr}_{T+\text{Rfn}(k,T)}(\varphi) : T + \text{Rfn}(k,T) \nvdash \varphi\}.$$
Then, by induction, $T + \bigcup\{X_k : k \leq n\} \vdash \text{Rfn}(n+1,T)$. Let $X = \bigcup\{X_k : k \in \mathbb{N}\}$. Then X is a (non-r.e.) set of true Π_1 sentences, whence $T + X$ is consistent, and $T + X \vdash \text{Rfn}_T^\omega$. Thus, by Theorem 2, $T + \text{Rfn}_T^\omega \nvdash \text{RFN}_T$. ∎

If T is Σ_1-sound then, by Theorem 4(a), $T + \text{Con}_T^\omega$ is a proper subtheory of $T + \text{Rfn}_T$. In our next result we show that if we restrict ourselves to Π_1 sentences, this is no longer true.

We write $S \dashv_{\Pi_1} S'$ to mean that S is a Π_1-*subtheory* of S', i.e., every Π_1 sentence provable in S is provable in S'.

Theorem 5. If $\text{PA} \dashv T$, then $T + \text{Rfn}_T \dashv_{\Pi_1} \text{PA} + \text{Con}_T^\omega$.

In the proof we use the following observation.

Lemma 3. $T \dashv_{\Pi_1} \text{PA} + \text{Con}_T$.

Proof. Let π be a Π_1 sentence such that $T \vdash \pi$. Then $\text{PA} \vdash \text{Pr}_T(\pi)$. Since $\neg \pi$ is Σ_1, we also have, $\text{PA} \vdash \neg \pi \to \text{Pr}_T(\neg \pi)$. It follows that $\text{PA} \vdash \neg \pi \to \neg \text{Con}_T$ and so $\text{PA} + \text{Con}_T \vdash \pi$. ∎

Lemma 4. Suppose $\text{PA} \dashv T$. For any n and any sentences $\varphi_0, \ldots, \varphi_{n-1}$,
$$\text{PA} \vdash \text{Pr}_T(\neg \bigwedge \{\text{Pr}_T(\varphi_i) \to \varphi_i : i < n\}) \to \neg \text{Con}(n+1, T).$$

Proof. We write $\text{Pr}(x)$ for $\text{Pr}_T(x)$ and Con^n for $\text{Con}(n,T)$. The statement holds for $n = 0$ (the conjunction of the empty set is T). Suppose it holds for n. Let $\theta_n := \neg \bigwedge \{\text{Pr}(\varphi_i) \to \varphi_i : i < n\}$.

Let $j < n+1$. Then
$$\text{PA} \vdash \text{Pr}(\theta_{n+1}) \to \text{Pr}(\varphi_j \to \neg \bigwedge \{\text{Pr}(\varphi_i) \to \varphi_i : i < n+1, i \neq j\}),$$
$$\to (\text{Pr}(\varphi_j) \to \text{Pr}(\neg \bigwedge \{\text{Pr}(\varphi_i) \to \varphi_i : i < n+1, i \neq j\})),$$
$$\to (\text{Pr}(\varphi_j) \to \neg \text{Con}^{n+1}), \text{ by ind. hyp.}$$

It follows that
(1) $\qquad \text{PA} \vdash \text{Pr}(\theta_{n+1}) \to (\bigvee \{\text{Pr}(\varphi_j) : j < n+1\} \to \neg \text{Con}^{n+1})$.

Clearly,
$$\text{PA} \vdash \theta_{n+1} \to \bigvee \{\text{Pr}(\varphi_j) : j < n+1\},$$
since the formula is a tautology. From this and (1) we get
$$\text{PA} \vdash \theta_{n+1} \wedge \text{Pr}(\theta_{n+1}) \to \neg \text{Con}^{n+1}.$$
It follows that

$$PA \vdash Pr(\theta_{n+1}) \rightarrow Pr(\theta_{n+1} \wedge Pr(\theta_{n+1})),$$
$$\rightarrow Pr(\neg Con^{n+1}),$$
$$\rightarrow \neg Con^{n+2},$$

as desired. ∎

Proof of Theorem 5. Let π be a Π_1 sentence and suppose $T + Rfn_T \vdash \pi$. Let n be such that $T_n \vdash \pi$, where $T_n = T + Pr(\varphi_0) \rightarrow \varphi_0 + ... + Pr(\varphi_{n-1}) \rightarrow \varphi_{n-1}$. By Lemma 3, $PA + Con_{T_n} \vdash \pi$. By Lemma 4, $PA \vdash Con(n+1, T) \rightarrow Con_{T_n}$. It follows that $PA + Con(n+1, T) \vdash \pi$ and so $PA + Con_T^\omega \vdash \pi$, as desired. ∎

For completeness we mention, but do not prove, that $PA + RFN_T$ is not a Π_1-subtheory of $T + Rfn_T^\omega$; for example, $PA + RFN_T \vdash Con_{T+Rfn_T^\omega}$.

§2. Irredundant axiomatizability.

A set X of sentences is *irredundant over* T if for every $\phi \in X$, $T + (X - \{\phi\}) \nvdash \phi$. An extension S of T is *irredundantly axiomatizable* (*i.a.*) *over* T if there is an axiomatization $T + X$ of S such that X is irredundant over T. In this case we shall also say that $T + X$ is *irredundant over* T. If S is a finite extension of T, then S is i.a. over T. A theory is *irredundantly axiomatizable* (*i.a.*) if it is i.a. over the empty theory (logic). If T is i.a. over a finite theory, then T is i.a.

Theorem 6. If $PA \dashv T$, then T is i.a.

Lemma 5. Suppose X is recursive and $S + X$ is not a finite extension of S. Then $S + X$ is i.a. over S iff there is a recursive function f such that for every conjunction χ of members of X, $S + X \vdash f(\chi)$ and $S \nvdash \chi \rightarrow f(\chi)$.

Proof. "If". Let f(n) be as assumed. Let $\varphi_0, \varphi_1, \varphi_2, ...$ be an effective enumeration of X. Let $\chi_n := \varphi_0 \wedge ... \wedge \varphi_n$. We may assume that $S \nvdash \varphi_0$. We effectively define sentences ψ_n in the following way. Let $\psi_0 := \varphi_0$. Suppose ψ_n has been defined and $S + X \vdash \psi_n$. We can then effectively find an m such that $S + \chi_m \vdash \psi_n$. Let $\psi_{n+1} := \chi_m \wedge f(\chi_m)$. Then $S + X \dashv\vdash S + \{\psi_n : n \in N\}$, $\vdash \psi_{n+1} \rightarrow \psi_n$, and $S \nvdash \psi_n \rightarrow \psi_{n+1}$ for every n. Next let $\theta_0 := \psi_0$ and $\theta_{n+1} := \psi_n \rightarrow \psi_{n+1}$. Again we have $S + X \dashv\vdash S + \{\theta_n : n \in N\}$. For every n, $S + \neg\theta_n$ is consistent. Also $\vdash \neg\theta_n \rightarrow \theta_k$ for every $k \neq n$. It follows that $S + \{\theta_k : k \neq n\} \nvdash \theta_n$. Thus, $S + \{\theta_n : n \in N\}$ is an axiomatization of $S + X$ which is irredundant over S.

"Only if". Let $S + Y$ be an axiomatization of $S + X$ which is irredundant over S. Let χ be a conjunction of members of X. Given χ, we can effectively find a conjunction ψ of members $\psi_0, ..., \psi_k$ of Y such that $S + \psi \vdash \chi$. Since $S + X$ is not finite over S, we can now effectively find a sentence $\theta \in$

$Y - \{\psi_0, \ldots, \psi_k\}$. Let $f(\chi) = \theta$; if n is not a conjunction of members of X, let $f(n) = 0$. Since $S + Y$ is irredundant over S, it follows that $f(n)$ is as desired. ■

Proof of Theorem 6. Let φ be as in Theorem 2.1 with $T = Q + \chi$. If $T \vdash \chi$, then $Q + \chi$ is consistent and so $Q + \chi \nvdash \varphi$. By Theorem 2.4, $PA + Con_{Q+\chi} \vdash \varphi$. Set $f(\chi) = \varphi$. Then $\vdash \chi \to f(\chi)$. Also, by Corollary 1.7, $T \vdash Con_{Q+\chi}$ and so $T \vdash f(\chi)$. The desired conclusion now follows from Lemma 5 with $S = \emptyset$ and $X = T$. ■

To prove the existence of non-i.a. theories we borrow the following lemma from recursion theory.

Lemma 6. There is a coinfinite r.e. set H such that for every recursive function h (such that $h(n) < h(n+1)$ for every n), there is a number m such that $\{k: h(m) < k \leq h(m+1)\} \subseteq H$. (It follows that H is not recursive.)

Theorem 7. There is a Π_1 (Σ_1) formula $\eta(x)$ such that $T + \{\eta(k): k \in N\}$ is not i.a. over T.

Proof. Let H be as in Lemma 6. By Theorem 3.3, there is a Π_1 (Σ_1) formula $\eta(x)$ numerating H in T and such that if $k \notin H$, then
$$T + \{\eta(m): m \neq k\} \nvdash \eta(k).$$
Let $S = T + \{\eta(k): k \in N\}$. Then, since H is coinfinite, S is not finite over T. Suppose S is i.a. over T. Let $\varphi_n := \eta(0) \wedge \ldots \wedge \eta(n)$. By Lemma 5, there is a recursive function f such that for every n, $S \vdash f(\varphi_n)$ and $T \nvdash \varphi_n \to f(\varphi_n)$. There is a recursive function g such that for every n, $T \vdash \varphi_{g(n)} \to f(\varphi_n)$. It follows that $T \nvdash \varphi_n \to \varphi_{g(n)}$. Let $h(0) = 0$ and $h(n+1) = g(h(n))$. Then for every n, $T \nvdash \varphi_{h(n)} \to \varphi_{h(n+1)}$. But $T \vdash \eta(k)$ for $k \in H$. It follows that $\{k: h(n) < k \leq h(n+1)\} \nsubseteq H$ for every n, contradicting Lemma 6. ■

Corollary 4. If T is finite, there is a Π_1 (Σ_1) formula $\eta(x)$ such that $T + \{\eta(k): k \in N\}$ is not i.a.

Let $S = T + \{\varphi_k: k \in N\}$. Suppose S is i.a. over T. By the proof of Lemma 5, there are conjunctions ψ_m of the sentences φ_k such that if $\theta_0 := \psi_0$, $\theta_{m+1} := \psi_m \to \psi_{m+1}$, then $T + \{\theta_k: k \in N\}$ is an axiomatization of S which is irredundant over T. However, irredundance has been obtained at the price of a slight increase in complexity: supposing that the sentences φ_k are Γ, it does not follow that this is true of the sentences θ_k. Thus, we may ask if irredundance can always be achieved without raising complexity. By our next result, the answer is negative.

Let us say that S is *irredundantly* Γ*-axiomatizable* (*i.* Γ*-a.*) over T, if there is an r.e. set $Z \subseteq \Gamma$ such that $T + Z$ is an axiomatization of S which is irredundant over T.

Theorem 8. If PA ⊣ T, there is a Π_n formula $\xi(x)$ such that $T + \{\xi(k): k \in N\}$ is i.a. over T but not i. Π_n-a. over T.

The proof of Theorem 8 uses methods which will be developed in Chapter 5; it is given at the end of §1 of that chapter.

Exercises for Chapter 4
1. (a) Show that
$$PA + \varphi + Rfn_S \vdash Rfn_{S+\varphi},$$
$$PA + \varphi + RFN_S \vdash RFN_{S+\varphi}.$$
   (b) Let
$$Rfn_S(\Gamma) = \{Pr_S(\varphi) \to \varphi: \varphi \text{ is } \Gamma\},$$
$$RFN_S(\Gamma) := \forall x(\Gamma(x) \wedge Pr_S(x) \to Tr_\Gamma(x)).$$
(i) Improve (a) by showing that
   if φ is Γ, then $PA + \varphi + Rfn_S(\Gamma^d) \vdash Rfn_{S+\varphi}(\Gamma^d)$,
   if φ is Γ, then $PA + \varphi + RFN_S(\Gamma^d) \vdash RFN_{S+\varphi}(\Gamma^d)$.
(ii) Show that
   if $Q \dashv S$, then $PA + Con_S \vdash RFN_S(\Pi_1)$,
   $PA + RFN_S(\Sigma_n) \vdash RFN_S(\Pi_{n+1})$.
(iii) Suppose PA ⊣ T. Show that
   if $X \subseteq \Gamma$ is r.e. and $T + X \vdash Rfn_T(\Gamma^d)$, then $T + X$ is inconsistent,
   if $X \subseteq \Gamma$ and $T + X \vdash RFN_T(\Gamma^d)$, then $T + X$ is inconsistent.
Define the sets $Rfn_S^\omega(\Gamma)$ and $RFN_S^\omega(\Gamma)$ in the natural way. Suppose S and T are true. Conclude that
$$T + Rfn_S^\omega \not\vdash RFN_T(\Sigma_1),$$
$$T + Rfn_S^\omega(\Sigma_n) \not\vdash Rfn_T(\Pi_n) \text{ for } n \geq 2,$$
$$T + RFN_S^\omega(\Pi_n) \not\vdash Rfn_T(\Sigma_n).$$

2. Suppose PA ⊣ T. Let
$$RFN_\tau = \{\forall x(\Gamma(x) \wedge Pr_\tau(x) \to Tr_\Gamma(x)): \Gamma \text{ arbitrary}\}.$$
Let φ be any sentence such that $T \not\vdash \varphi$. Show that there is a PR binumeration $\tau(x)$ of T such that $T + RFN_\tau \not\vdash \varphi$.

3. Suppose PA ⊣ T and T is Σ_1-sound.

(a) Show that $T + \text{Rfn}_T(\Gamma)$ is not essentially infinite over T.

(b) Let S be such that $T + \text{Rfn}_T(\Sigma_1) \dashv S \dashv T + \text{Rfn}_T$. Show that S is infinite over T. [Hint: Use (the proof of) Theorem 5 and Theorem 2.4.]

4. (a) Suppose the formula $\alpha(x)$ is such that for every φ,
if $T \vdash \varphi$, then $T \vdash \alpha(\varphi)$.
Show that there is a sentence ψ such that $T \nvdash \alpha(\psi) \to \psi$. [Hint: Use Exercise 2.11.]

(b) Suppose there is a formula $\alpha(x)$ such that for every φ,
if $\vdash \varphi$, then $T \vdash \alpha(\varphi)$,
$T \vdash \alpha(\varphi) \to \varphi$.
Show that T is not finitely axiomatizable. (This also follows by the proof of Theorem 1 with $S = \emptyset$.)

5. T is *reducible to* S if there is a recursive function g such that for all sentences φ, (i) $T \vdash g(\varphi)$ and (ii) if $T \vdash \varphi$, then $S \vdash g(\varphi) \to \varphi$. If T is a finite extension of S, $T = S + \theta$, then T is reducible to S: let $g(\varphi) = \theta$ for every φ. Prove the following result, a strengthening of Theorem 1: If $T \vdash \text{Rfn}_S$, then T is not reducible to S. [Hint: Suppose T is reducible to S and let g be the relevant recursive function. Let ψ be such that
$$Q \vdash \psi \leftrightarrow \neg \text{Pr}_S(g(\psi) \to \psi).$$
Show that $T \vdash \psi$ and $Q \vdash \neg \psi$.]

6. We write $\text{Pr}(x)$, Con^n, Rfn, $\text{Rfn}(\Gamma)$ for $\text{Pr}_T(x)$, $\text{Con}(n,T)$, Rfn_T, $\text{Rfn}_T(\Gamma)$, respectively. Let $\varphi_0, \varphi_1, \ldots$ be an enumeration of all sentences. Let $\chi_n := \bigwedge\{\text{Pr}(\varphi_i) \to \varphi_i : i < n\}$.

(a) For each n and φ, let $\beta_n[\varphi]$ be defined as follows: $\beta_0[\varphi] := \varphi$, $\beta_{n+1}[\varphi] := \varphi \vee \text{Pr}(\beta_n[\varphi])$.

Show that
(1) $\text{PA} \vdash (\chi_n \to \varphi) \wedge \text{Pr}(\chi_n \to \varphi) \to \beta_n[\varphi]$.
(2) $\text{PA} \vdash \text{Pr}(\chi_n \to \varphi) \to \text{Pr}(\beta_n[\varphi])$,
$\text{PA} \vdash (\chi_n \to \varphi) \wedge \text{Pr}(\chi_n \to \varphi) \to (\bigwedge\{\text{Pr}(\beta_i[\varphi]) \to \beta_i[\varphi] : i < n\} \to \varphi)$.

Suppose $\Gamma \neq \Pi_1$. Conclude that if φ is Γ and $T + \text{Rfn} \vdash \varphi$, then $\text{PA} + \text{Rfn}(\Gamma) \vdash \varphi$. (Lemma 4 can be obtained from (2) by replacing φ by $\bot$.) [Hint: The proof of (1) is similar to that of Lemma 4. $\beta_n[\varphi] \to (\bigwedge\{\text{Pr}(\beta_i[\varphi]) \to \beta_i[\varphi] : i < n\} \to \varphi)$ is a tautology.]

(b) Let $\text{Pr}^1(x) := \text{Pr}(x)$, $\text{Pr}^{n+1}(x) := \text{Pr}(\text{Pr}^n(\dot{x}))$. Clearly,
$$T + \text{Pr}^n(\bot) \to \text{Pr}^{n-1}(\bot) + \ldots + \text{Pr}(\bot) \to \bot \vdash \text{Con}^n.$$
Thus, Con^n follows from n instances of Rfn in T. Show that

$$PA \vdash Pr(\chi_n \to Con^{n+1}) \to \neg Con^{n+1}.$$
Conclude that if T is Σ_1-sound, then Con^{n+1} is not derivable from n instances of Rfn in T.

7. (a) Suppose $S \nvdash \varphi$ and $S + \neg \varphi + Z$ is non-i.a. over $S + \neg \varphi$. Show that $S + \{\varphi \vee \psi : \psi \in Z\}$ is non-i.a. over S.
    (b) Suppose $T \dashv_p T'$. Show that
(i) there is a theory S such that $T \dashv S \dashv T'$ and S is not i.a. over T,
(ii) if T is finitely axiomatizable, there is a theory S such that $T \dashv S \dashv T'$ and S is not i.a.

8. T is *effectively e.i. over* S if $T \vdash S$ and there is a recursive function f such that for every φ, (i) $T \vdash f(\varphi)$ and (ii) if $S + \varphi \vdash f(\varphi)$, then $T \vdash \neg \varphi$ (and so if $T \dashv S + \varphi$, then $S + \varphi$ is inconsistent).
    (a) Show that if T is effectively e.i. over S, then T is i.a. over S.
    (b) Suppose $Q \dashv S$. Show that if $T \vdash Rfn_S$, then T is effectively e.i. over S. (Compare Exercise 5.11.)

9. Suppose $PA \dashv T$. Let X and Y be any r.e. sets of Γ sentences such that if $\varphi \in X$ and $\psi \in Y$, then $T \vdash \varphi \to \psi$. Show that there is a Γ sentence θ such that if $\varphi \in X$ and $\psi \in Y$, then $T \vdash \varphi \to \theta$ and $T \vdash \theta \to \psi$. [Hint: Suppose $\Gamma = \Pi_n$. Suppose X and Y are primitive recursive. Let $\xi(x)$ and $\eta(x)$ be PR binumerations of X and Y. Let $\theta :=$
$$\forall x(\eta(x) \wedge \forall y \leqslant x(\xi(y) \to \neg Tr_{\Pi_n}(y)) \to Tr_{\Pi_n}(x)).]$$

10. Suppose $PA \dashv T$ and T is Σ_1-sound.
    (a) Show that there is a Π_1 formula $\beta(x)$ such that for every m, $T + \beta(m)$ is consistent and
$$T + \beta(m) \vdash Con_{T+\beta(m+1)}.$$
(Note that $T + \beta(m) \vdash \beta(m+1)$.) [Hint: Let the primitive recursive function f be defined (in T and in the real world) as follows; we assume that $\delta(x)$ is a PR formula:
$$f(\delta,\xi,0) = 0,$$
$$f(\delta,\xi,n+1) = m \text{ if } m > f(\delta,\xi,n),$$
$$\forall z \leqslant m \neg \delta(z),$$
n is a proof in T of $\neg \xi(\delta,m)$,
if there is such a number m,
$$= f(\delta,\xi,n), \text{ otherwise.}$$
If the value of f(k,m,n) is not determined by these conditions, it is irrele-

vant and we may set $f(k,m,n) = 0$.
  Next let $\gamma(z,x)$ be such that
$$PA \vdash \gamma(z,x) \leftrightarrow \forall y(f(z,\gamma,y) \leq x).$$
Let $g(k,s) = f(k,\gamma,s)$.
  Claim. If $\exists x \delta(x)$ is true, then for every n, $g(\delta,n) = 0$.
Proof. Let k be the least number such that $\delta(k)$ is true. Then for every n, $g(\delta,n) \leq k$. Thus, if the claim is false, there is a largest n such that $g(\delta,n) \neq g(\delta,n+1)$. Let $m = g(\delta,n+1)$. Then n is a proof of $\neg\gamma(\delta,m)$. It follows that $\neg\forall y(g(\delta,y) \leq m)$ is provable and so is true, a contradiction.
  Let $\delta'(x)$ be a PR formula such that
$$PA \vdash \exists x \delta'(x) \leftrightarrow Pr_T(\neg\forall y(g(\delta',y) = 0)).$$
Let $\beta(x) := \forall y(g(\delta',y) \leq x).]$
  (b) Show that $T + RFN_T(\Sigma_1) \vdash \beta(0)$, that $T + Rfn_T \dashv_{\Pi_1} T + \beta(0)$, and that $T + Rfn_T \nvdash \beta(0)$.
  (c) Show that with each rational number $a \geq 0$, we can effectively associate a Π_1 sentence θ_a such that $T + \theta_a$ is consistent and if $a < b$, then $T + \theta_a \vdash Con_{T+\theta_b}$. [Hint: Define a function g in much the same way as in case (a) except that g may, in a sense, take rational numbers ≥ 0 as values.]

Notes for Chapter 4
Theorems 1 and 2 are due to Kreisel and Lévy (1968). The formula $Pr_{S,\Gamma}(x)$ and the present formulation of the proof of Theorem 2 are due to Smoryński (1981b). Corollary 1(a) is due to Montague (1961) and Rabin (1961). What we have called the uniform reflection principle RFN_S is not quite what is usually referred to by that term, but for theories containing PA the difference is negligible. Theorem 3 is due to Lindström (1984a). Corollary 2 is due to Kreisel and Lévy (1968). Theorem 4(b) is a weak form of a result of Feferman (1962). For (partial) improvements of Theorems 1, 2, 4 and Corollaries 1, 2, see Exercise 1. Theorem 5 is due to Goryachev (1986) (with a different proof; cf. also Beklemishev (1995) and Lindström (1996)).
  More information on (transfinite) interations of consistency statements and reflection principles, a rather technical subject which falls outside the scope of this book, can be found in Feferman (1962) and Beklemishev (1995).
  What we have called an irredundant axiomatization is usually called an independent axiomatization. Theorem 6 is due to Montague and Tarski (1957). Lemma 5 is due to Tarski (cf. Montague and Tarski (1957)). For a

proof of the existence of an r.e. set as described in Lemma 6, a so called *hypersimple* set, see Soare (1987). The idea of using hypersimple sets to construct non-i.a. theories is due to Kreisel (1957). Theorem 7 is related to a result of Pour-El (1968) and Corollary 4 is Pour-El's result restricted to theories in L_A. Theorem 8 is new; Theorem 8 with Π_n replaced by Σ_n and restricted to Σ_n-sound theories is also true but seems to require a quite different proof.

Exercise 3(b) is due to Beklemishev (1997). Exercise 4 is due to Montague (1963). Exercise 5 is due to Kreisel and Lévy (1968). Exercise 6(a) is due to Beklemishev (1997) (with a different proof); Exercise 6(b) is due to Artemov (1979) and Boolos (1979a), (1993) (with different proofs); cf. also Lindström (1996); the suggested proofs were inspired by Beklemishev (1997). Exercise 10(a) was proved by Harvey Friedman, Smoryński, and Solovay, independently, answering a question of Haim Gaifman; for a different proof, due to Friedman, see Smoryński (1985), p. 179. Exercise 10(c) is due to Alex Wilkie (with a different proof); see Simmons (1988). The present proofs of (a) and (c) can be modified to yield much stronger conclusions.

5. PARTIAL CONSERVATIVITY

A sentence φ is Γ-*conservative over* T if for every Γ sentence θ, if $T + \varphi \vdash \theta$, then $T \vdash \theta$. In this chapter we study this phenomenon for its own sake. Results on Γ-conservativity are, however, also very useful in many contexts, in particular in connection with interpretability (see Chapters 6 and 7).

Our task in this chapter is to develop general methods for constructing partially conservative sentences satisfying additional conditions such as being nonprovable in a given theory.

We assume throughout that PA ⊣ T. The results of this chapter do not depend on the assumption that T is reflexive.

§1. Partial conservativity. A first example of a Π_1-conservative sentence is given in the following:

Theorem 1. $\neg \mathrm{Con}_T$ is Π_1-conservative over T.

Proof. Suppose θ is Π_1 and
(1) $\quad T + \neg \mathrm{Con}_T \vdash \theta$.
From (1) we get $\mathrm{PA} \vdash \mathrm{Pr}_T(\neg\theta) \to \mathrm{Pr}_T(\mathrm{Con}_T)$, whence
(2) $\quad \mathrm{PA} \vdash \mathrm{Pr}_T(\neg\theta) \to \neg \mathrm{Con}_{T + \neg \mathrm{Con}_T}$.
By provable Σ_1-completeness,
(3) $\quad \mathrm{PA} \vdash \neg\theta \to \mathrm{Pr}_T(\neg\theta)$.
By Corollary 2.2,
(4) $\quad \mathrm{PA} + \mathrm{Con}_T \vdash \mathrm{Con}_{T + \neg \mathrm{Con}_T}$.
Combining (2), (3), (4) we get $\mathrm{PA} \vdash \neg\theta \to \neg \mathrm{Con}_T$ and so by (1), $T \vdash \theta$. ∎

By Corollary 2.5, Theorem 1 provides us with an example of a (Σ_1) sentence φ which is Π_1-conservative over T and nontrivially so, i.e., such that $T \nvdash \varphi$, even if T is not Σ_1-sound.

If φ is Γ-conservative over T and ψ is Γ^d, then clearly φ is Γ-conservative over $T + \psi$. Suppose T is Σ_1-sound. Then if π is Π_1, then π is Σ_1-conservative over T iff π is true iff $T + \pi$ is consistent. Also note that for every sentence φ, either φ or $\neg\varphi$ is Σ_1-conservative over T.

Let us now try to construct a sentence φ which is nontrivially Γ-conservative over T. Thus, given that
(1) $\quad T + \varphi \vdash \theta$,

§1. Partial Conservativity

where θ is Γ, we want to be able to conclude that $T \vdash \theta$. This follows if (1) implies that

(2) $T + \neg\theta \vdash \varphi$.

The natural way to ensure that (1) implies (2) is to let φ be a sentence saying of itself that there is a false Γ sentence (namely θ) which φ implies in T. Thus, let φ be such that

(3) $PA \vdash \varphi \leftrightarrow \exists u(\Gamma(u) \wedge Pr_{T+\varphi}(u) \wedge \neg Tr_\Gamma(u))$,

where $\Gamma(x)$ is a PR binumeration of the set of Γ sentences. Then (1) implies (2).

It is, however, not generally true that $T \nvdash \varphi$. This holds if T is true, since φ is then false. But, for example, $T + \neg Con_T \vdash \varphi$, and so if $T \vdash \neg Con_T$, then $T \vdash \varphi$. To prevent this from happening, we redefine φ as follows: let φ be such that

$PA \vdash \varphi \leftrightarrow \exists y \exists u v \leqslant y(\Gamma(u) \wedge Prf_{T+\varphi}(u,v) \wedge \neg Tr_\Gamma(u) \wedge \forall z \leqslant y \neg Prf_T(\varphi,z))$.

Then $T \nvdash \varphi$ and φ is Γ-conservative over T. Also, if $\Gamma = \Pi_n$, then φ is Σ_n which is optimal; in fact, this is the sentence used in the proof of Theorem 2(a), below, for $\Gamma = \Sigma_n$.

From our present point of view the proof of Theorem 4.2 with $S = T$ can be understood as follows (see the remarks following Corollary 4.1). Let ψ be as in that proof. It is sufficient to show that $\neg\psi$ is Γ^d-conservative over T; in fact, that is exactly what is done in the proof of Theorem 4.2. This also follows from the fact that (3) with φ replaced by $\neg\psi$ and Γ by Γ^d is true.

Let $[\Gamma]_S(x,y) :=$
$\forall u v \leqslant y(\Gamma(u) \wedge Prf_{S+x}(u,v) \rightarrow Tr_\Gamma(u))$.

This formula is constructed to yield the following:

Lemma 1. $[\Gamma]_T(x,y)$ is a Γ formula such that
(i) $PA \vdash [\Gamma]_T(x,y) \wedge z \leqslant y \rightarrow [\Gamma]_T(x,z)$,
(ii) $T + \varphi \vdash [\Gamma]_T(\varphi,m)$ for all φ and m,
(iii) if ψ is Γ and $T + \varphi \vdash \psi$, there is a q such that $PA + [\Gamma]_T(\varphi,q) \vdash \psi$.

Proof. (i) is clear. (ii) Let $\theta_0,...,\theta_k$ be all Γ sentences $\leqslant$ m provable in $T + \varphi$ whose proofs in $T + \varphi$ are $\leqslant$ m. Then
$PA \vdash \forall u v \leqslant m(\Gamma(u) \wedge Prf_{T+\varphi}(u,v) \rightarrow u = \theta_0 \vee ... \vee u = \theta_k)$.
Also clearly, by Fact 10(a) (ii),
$T + \varphi \vdash u = \theta_0 \vee ... \vee u = \theta_k \rightarrow Tr_\Gamma(u)$.
It follows that $T + \varphi \vdash [\Gamma]_T(\varphi, m)$.

(iii) Suppose ψ is Γ and $T + \varphi \vdash \psi$. Let q be a proof of ψ in $T + \varphi$. Then $\psi < q$. It follows that $PA + [\Gamma]_T(\varphi,q) \vdash Tr_\Gamma(\psi)$ and so $PA + [\Gamma]_T(\varphi,q) \vdash \psi$. ∎

S is a Γ-*subtheory* of T, S ⊣$_Γ$ T, if every Γ sentence provable in S is provable in T. In what follows we write [Γ](x,y) for [Γ]$_T$(x,y).

Lemma 2. Suppose χ(x,y) is Γ. There is then a Γ formula ξ(x) such that for all k and m,
(i) T + ξ(k) ⊢ χ(k,m),
(ii) T + ξ(k) ⊣$_{Γd}$ T + {χ(k,q): q ∈ N}.

Proof. *Case 1.* Γ = Π_n. Let ξ(x) be such that
(1)   PA ⊢ ξ(k) ↔ ∀y([Σ_n](ξ(k),y) → χ(k,y)).
Then (i) follows from Lemma 1(ii) and (1). To prove (ii), suppose ψ is Σ_n and T + ξ(k) ⊢ ψ. By Lemma 1(iii), there is a q such that
    PA + [Σ_n](ξ(k),q) ⊢ ψ.
Hence, by Lemma 1(i),
    PA + ∀y≤qχ(k,y) + ¬ψ ⊢ ∀y([Σ_n](ξ(k),y) → χ(k,y))
and so, by (1),
    PA + ∀y≤qχ(k,y) + ¬ψ ⊢ ξ(k).
But then, since T + ξ(k) ⊢ ψ, it follows that T + {χ(k,q): q ∈ N} ⊢ ψ, as desired.
Case 2. Γ = Σ_n. Let ξ(x) be such that
    PA ⊢ ξ(k) ↔ ∃y(¬[Π_n](ξ(k),y) ∧ ∀z≤yχ(k,z)).
The proof that ξ(x) is as claimed is then almost the same as in Case 1. ∎

From Lemma 2 we derive the following result on numerations of r.e. sets.

Lemma 3. Let X be an r.e. set. There is then a Γ formula ξ(x) such that
(i) if k ∈ X, then T ⊢ ξ(k),
(ii) if k ∉ X, then ¬ξ(k) is Γ-conservative over T.

Proof. Let ρ(x,y) be a PR formula such that X = {k: ∃mPA ⊢ ρ(k,m)}. By Lemma 2, there is a Γ formula ξ(x) (the negation of the formula ξ(x) of Lemma 2) such that
    T + ¬ξ(k) ⊢ ¬ρ(k,m),
    T + ¬ξ(k) ⊣$_Γ$ T + {¬ρ(k,q): q ∈ N}.
ξ(x) is then as desired. ∎

For extensions of PA Lemma 3 implies Theorem 3.1.

We can now prove our first general theorem on the existence of nontrivially partially conservative sentences.

§1. Partial Conservativity

Theorem 2. (a) There is a Γ^d sentence φ such that φ is Γ-conservative over T and $T \nvdash \varphi$.

(b) If X is r.e. and monoconsistent with T, there is a Γ^d sentence φ such that φ is Γ-conservative over T and $\varphi \notin X$.

Proof. (a) is the special case of (b) where $X = Th(T)$. ♦

(b) Let $\xi(x)$ be as in Lemma 3 and let φ be such that $PA \vdash \varphi \leftrightarrow \neg\xi(\varphi)$. If $\varphi \in X$, then $T \vdash \xi(\varphi)$ and so $T \vdash \neg\varphi$, which is impossible. Thus, $\varphi \notin X$ and so φ is Γ^d-conservative over T. ∎

Of course, the Γ^d sentence mentioned in Theorem 2(a) is not Γ^T (compare Corollary 2.6).

The following result is a natural strengthening of Theorem 2.

Theorem 3. (a) There is a Γ sentence φ such that φ is Γ^d-conservative over T and $\neg\varphi$ is Γ-conservative over T.

(b) If X is r.e. and monoconsistent with T, there is a Γ sentence φ such that φ is Γ^d-conservative over T, $\neg\varphi$ is Γ-conservative over T, and $\varphi, \neg\varphi \notin X$.

We derive Theorem 3 from:

Lemma 4. Suppose $\chi_0(x,y)$ is Γ^d and $\chi_1(x,y)$ is Γ. Then there is a Γ formula $\xi(x)$ such that for all k and m and $i = 0, 1$,
(i) $T + \xi^i(k) \vdash \forall y \leqslant m \chi_i(k,y) \to \chi_{1-i}(k,m)$,
(ii) if ψ is Γ^d and $T + \xi^i(k) \vdash \psi^i$, then $T + \{\chi_{1-i}(k,q) : q \in N\} \vdash \psi^i$.

Proof. We need only prove this for $\Gamma = \Sigma_n$. Let $\xi(x)$ be such that
(1) $\quad PA \vdash \xi(k) \leftrightarrow \exists y((\neg[\Pi_n](\xi(k),y) \vee \neg\chi_0(k,y)) \wedge$
$$\forall z < y ([\Sigma_n](\neg\xi(k),z) \wedge \chi_1(k,z))).$$
We verify (i) and (ii) for $i = 0$ and leave the case $i = 1$ to the reader.

(i) By Lemma 1 (ii),
$$T + \xi(k) \vdash \neg[\Pi_n](\xi(k),y) \to y > m.$$
It follows that
$$T + \xi(k) + \forall y \leqslant m \chi_0(k,y) \vdash (\neg[\Pi_n](\xi(k),y) \vee \neg\chi_0(k,y)) \to y > m.$$
But then, by (1),
$$T + \xi(k) + \forall y \leqslant m \chi_0(k,y) \vdash \chi_1(k,m),$$
as desired.

(ii) Suppose ψ is Π_n and
(2) $\quad T + \xi(k) \vdash \psi$.
By Lemma 1(iii), there is a q such that $T + [\Pi_n](\xi(k),q) \vdash \psi$ and so

(3) $T + \neg\psi \vdash \neg[\Pi_n](\xi(k),q)$.

By Lemma 1(ii),
$$T + \neg\xi(k) \vdash [\Sigma_n](\neg\xi(k),q).$$
From this, (1), and (3) it follows that
$$T + \neg\psi + \neg\xi(k) + \forall y \leqslant q \chi_1(k,y) \vdash \xi(k)$$
and so
$$T + \neg\psi + \forall y \leqslant q \chi_1(k,y) \vdash \xi(k).$$
But then, by (2), $T + \forall y \leqslant q \chi_1(k,y) \vdash \psi$, as desired. ∎

Proof of Theorem 3. (a) is a special case of (b). ♦

(b) Let $\rho_i(x,y)$, $i = 0, 1$, be PR binumerations of relations $R_i(k,m)$ such that $X = \{k : \exists m R_0(k,m)\}$ and $\{\varphi : \neg\varphi \in X\} = \{k : \exists m R_1(k,m)\}$. Let $\xi(x)$ be as in Lemma 4 with $\chi_i(x,y) := \neg\rho_{1-i}(x,y)$. Let φ be such that $PA \vdash \varphi \leftrightarrow \xi(\varphi)$. Suppose $\varphi \in X$ or $\neg\varphi \in X$. Let m be the least number such that $R_0(\varphi,m)$ or $R_1(\varphi,m)$. Suppose $R_i(\varphi,m)$. Then not $R_{1-i}(\varphi,n)$ for $n \leqslant m$. (We may assume that $R_0(k,n)$ implies not $R_1(k,n)$.) But then, by Lemma 4(i), $T \vdash \neg\xi^i(\varphi)$, whence $T \vdash \neg\varphi^i$. But this is impossible, since $\varphi^i \in X$. It follows that $\varphi, \neg\varphi \notin X$. But then, by Lemma 4(ii), φ is Γ^d-conservative over T and $\neg\varphi$ is Γ-conservative over T. ∎

In §2 we give an alternative proof of Theorem 3(a).

Our next result is related to Theorem 4.3; it will be used several times, in some cases indirectly, in Chapters 6 and 7.

S is a Γ-*conservative extension* of T if $T \dashv S \dashv_\Gamma T$. By Theorems 4.4(a) and 4.5, $T + \text{Rfn}_T$ is a Π_1-conservative extension of $PA + \text{Con}_T^\omega$.

Theorem 4. (a) Let X be an r.e. set of Γ sentences. There is then a Γ sentence θ such that $T + \theta$ is a Γ^d-conservative extension of $T + X$.

(b) Let $\gamma(x,y)$ be any Γ formula. There is then a Γ formula $\eta(x)$ such that for every k, $T + \eta(k)$ is a Γ^d-conservative extension of $T + \{\gamma(k,m) : m \in N\}$.

Proof. (a) By Craig's theorem, we may assume that X is primitive recursive. Let $\eta(x)$ be a PR binumeration of X. Then for every q,

(1) $PA + X \vdash \eta(q) \to \text{Tr}_\Gamma(q)$.

By Lemma 2 with $\chi(x,y) := \eta(y) \to \text{Tr}_\Gamma(y)$, there is a Γ sentence θ such that for all φ,

(2) $T + \theta \vdash \eta(\varphi) \to \text{Tr}_\Gamma(\varphi)$,

(3) $T + \theta \dashv_{\Gamma d} T + \{\eta(q) \to \text{Tr}_\Gamma(q) : q \in N\}$.

From (2) it follows that $T + \theta \vdash X$ and from (1) and (3) it follows that $T + \theta \dashv_{\Gamma d} T + X$. ♦

(b) Left to the reader. ∎

§1. Partial Conservativity

So far there has been no indication that the properties of Σ_n and Π_n, $n > 1$, in terms of partial conservativity may be different, but we shall now show that they are.

Let ψ_0 and ψ_1 be Γ sentences. If
(*) $\quad T \vdash \psi_0 \vee \psi_1$,
then, trivially,
(**) $\quad \psi_i$ is Γ^d-conservative over $T + \neg\psi_{1-i}$, $i = 0, 1$.
If $\Gamma = \Pi_n$, the converse of this is true. This follows from our next:

Lemma 5. Let ψ_0 and ψ_1 be any Π_n sentences. There are then Π_n sentences θ_0 and θ_1 such that
(i) $\quad PA \vdash \theta_0 \vee \theta_1$,
(ii) $\quad PA \vdash \psi_i \rightarrow \theta_i$, $i = 0, 1$,
(iii) $\quad PA \vdash \theta_0 \wedge \theta_1 \rightarrow \psi_0 \wedge \psi_1$.

Proof. By Fact 5, we may assume that $\psi_i := \forall x \delta_i(x)$, where $\delta_i(x)$ is Σ_{n-1}. Let $\theta_i :=$
$$\forall x (\neg \delta_i(x) \rightarrow \exists y < x + i \neg \delta_{1-i}(y)).$$
Then (i), (ii), (iii) are easily verified (cf. Lemma 1.3). ∎

Suppose $\Gamma = \Pi_n$. From (ii) and (iii) of Lemma 5 it follows that $T + \neg\psi_i + \psi_{1-i} \vdash \neg\theta_i$. Hence, assuming (**), $T + \neg\psi_i \vdash \neg\theta_i$. It follows that $T \vdash \theta_0 \vee \theta_1 \rightarrow \psi_0 \vee \psi_1$ and so, by Lemma 5(i), we get (*).

We now prove that if $\Gamma = \Sigma_n$, then (**) does not imply (*).

Theorem 5. (a) There are Σ_n sentences σ_0, σ_1 such that
(i) $\quad T \vdash \neg(\sigma_0 \wedge \sigma_1)$,
(ii) $\quad T \nvdash \sigma_0 \vee \sigma_1$,
(iii) $\quad \sigma_i$ is Π_n-conservative over $T + \neg\sigma_{1-i}$, $i = 0, 1$.

(b) Suppose X is r.e. and monoconsistent with T. Then there are Σ_n sentences σ_0, σ_1 such that (i) and (iii) hold and
(iv) $\quad \sigma_0 \vee \sigma_1 \notin X$.

We derive this theorem from:

Lemma 6. Let X be an r.e. set. There are then Σ_n formulas $\xi_0(x)$ and $\xi_1(x)$ such that for $i = 0, 1$,
(i) $\quad T \vdash \neg(\xi_0(x) \wedge \xi_1(x))$,
(ii) $\quad$ if $k \in X$, then $T \vdash \neg \xi_i(k)$,
(iii) $\quad$ if $k \notin X$, then $\xi_i(k)$ is Π_n-conservative over $T + \neg\xi_{1-i}(k)$.

Proof. Let $\rho(x,y)$ be a PR formula such that $X = \{k: \exists m PA \vdash \rho(k,m)\}$. For $i = 0$, 1, let $\xi_i(x), \chi_i(x), \delta_i(x,y)$ be, respectively, Σ_n, Σ_n, and Π_{n-1} formulas such that
(1) $PA \vdash \chi_i(k) \leftrightarrow \exists y(\neg[\Pi_n](\xi_i(k),y) \wedge \forall z \leqslant y \neg \rho(k,z))$,
(2) $PA \vdash \chi_i(x) \leftrightarrow \exists y \delta_i(x,y)$,
   $\xi_i(x) := \exists y(\delta_i(x,y) \wedge \forall z < y+i \neg \delta_{1-i}(x,z))$.
This application of (double) self-reference is more complicated than any we have encountered so far and it requires some thought to see that it is admissible. But in view of Fact 5 it is.

(i) is then clear. To prove (ii), suppose $k \in X$. Let m be such that $PA \vdash \rho(k,m)$. By Lemma 1 (ii),
$$T + \xi_i(k) \vdash \neg[\Pi_n](\xi_i(k),y) \rightarrow m < y.$$
So, by (1),
(3) $T + \xi_i(k) \vdash \neg \chi_i(k)$.
Also, by (2), $PA \vdash \xi_i(x) \rightarrow \chi_i(x)$. Now (ii) follows from this and (3).

Finally, to prove (iii), suppose $k \notin X$. Now suppose ψ is Π_n and
$$T + \neg \xi_{1-i}(k) + \xi_i(k) \vdash \psi.$$
By (i), it follows that
(4) $T + \xi_i(k) \vdash \psi$.
But then, by Lemma 1(iii), there is a q such that $T + [\Pi_n](\xi_i(k),q) \vdash \psi$. Also $T \vdash \neg \rho(k,m)$ for all m. By (1), it now follows that $T + \neg \psi \vdash \chi_i(k)$. Thus, by (2), $T + \neg \psi \vdash \exists y \delta_i(k,y)$. But then
$$T + \neg \psi + \neg \xi_{1-i}(k) \vdash \xi_i(k).$$
Combining this with (4) we get $T + \neg \xi_{1-i}(k) \vdash \psi$. This proves (iii). ∎

Proof of Theorem 5. (a) follows from (b). ◆

(b) We may assume that if $\psi \in X$ and $T \vdash \psi \rightarrow \theta$, then $\theta \in X$. Let $\xi_i(x)$ be as in Lemma 6. Let φ be such that
$$PA \vdash \varphi \leftrightarrow \xi_0(\varphi) \vee \xi_1(\varphi).$$
Set $\sigma_i := \xi_i(\varphi)$. If $\varphi \in X$, then, by Lemma 6(ii), $T \vdash \neg \xi_i(\varphi)$ for $i = 0, 1$, and so $T \vdash \neg \varphi$, impossible. Thus, $\varphi \notin X$ and so (iv) holds. (i) and (iii) follow from Lemma 6(i) and (iii), respectively. ∎

We can now partially improve Corollary 2.6 as follows:

Corollary 1. There are Σ_n sentences σ_0, σ_1, such that $T \vdash \sigma_0 \rightarrow \neg \sigma_1$ and there is no Δ_n sentence φ for which $T \vdash \sigma_0 \rightarrow \varphi$ and $T \vdash \varphi \rightarrow \neg \sigma_1$.

Proof. Let σ_0, σ_1 be as in Theorem 5(a). Suppose φ is Δ_n, $T \vdash \sigma_0 \rightarrow \varphi$, and $T \vdash \varphi \rightarrow \neg \sigma_1$. Then $T \vdash \neg \sigma_1 \rightarrow \varphi$ and $T \vdash \neg \sigma_0 \rightarrow \neg \varphi$ and so $T \vdash \sigma_0 \vee \sigma_1$, a contradiction. ∎

Let $\mathrm{Cons}(\Gamma, T)$ be the set of sentences Γ-conservative over T. It is clear from the definition of $\text{Cons}(\Gamma,T)$ that it is a Π_2^0 set. We now show that this classification is correct.

Our next lemma follows at once from Lemma 3.2(b) but has a simpler direct proof which we leave to the reader.

Lemma 7. Let $R(k,m)$ be any r.e. relation. There are then formulas $\rho_0(x,y)$ and $\rho_1(x,y)$ such that $\rho_0(x,y)$ is Σ_1, $\rho_1(x,y)$ is Π_1, $\rho_0(x,y)$ numerates $R(k,m)$ in T, $\text{PA} \vdash \rho_0(k,m) \to \rho_1(k,m)$, and if not $R(k,m)$, then $T \nvdash \rho_1(k,m)$.

Theorem 6. (a) $\text{Cons}(\Gamma,T)$ is a complete Π_2^0 set.
(b) If $\Gamma \neq \Sigma_1$, then $\Gamma^d \cap \text{Cons}(\Gamma,T)$ is a complete Π_2^0 set.

Proof. Let X be any Π_2^0 set and let $R(k,m)$ be an r.e. relation such that $X = \{k: \forall m R(k,m)\}$. Let $\rho(x,y)$ be a formula numerating $R(k,m)$ in T, which is Σ_1 if $\Gamma = \Sigma_n$ and Π_1 if $\Gamma = \Pi_n$. Let $\xi(x)$ be as in (the proof of) Lemma 2 with Γ replaced by Γ^d and $\chi(x,y) := \rho(x,y)$. Then
(1) $T + \xi(k) \vdash \rho(k,m)$,
(2) $T + \xi(k) \dashv_\Gamma T + \{\rho(k,q): q \in N\}$.
To prove (a) it is now sufficient to show that
(3) $k \in X$ iff $\xi(k) \in \text{Cons}(\Gamma,T)$.
If $k \in X$, then $T \vdash \rho(k,q)$ for every q and so, by (2), $\xi(k) \in \text{Cons}(\Gamma,T)$. If $k \notin X$, there is an m such that $T \nvdash \rho(k,m)$ and so, by (1), $\xi(k) \notin \text{Cons}(\Gamma,T)$ (in fact, $\xi(k)$ is not Σ_1- or not Π_1-conservative over T, as the case may be). Thus, (3) holds. This proves (a).

If Γ is Σ_n or Π_n with $n \geq 2$, then $\xi(x)$ is Γ^d as claimed in (b). Finally, suppose $\Gamma = \Pi_1$. Let $\rho_0(x,y)$ and $\rho_1(x,y)$ be as in Lemma 7. Let $\rho(x,y) := \rho_0(x,y)$. Then $\xi(x)$ is Σ_1. By Lemma 7, $\xi(k) \notin \text{Cons}(\Pi_1,T)$ if $k \notin X$. Thus, (b) holds in this case, too. ∎

Suppose T is Σ_1-sound and θ is Π_1. Then θ is Σ_1-conservative over T iff θ is true. Thus, $\Pi_1 \cap \text{Cons}(\Sigma_1,T)$ is Π_1^0.

We conclude this section with a proof of Theorem 4.8. We derive this result from the following lemma; a refinement of this lemma (for n = 1) will be proved in Chapter 7 (Lemma 7.23).

Lemma 8. There is a Π_n formula $\xi(x)$ such that for every k,
(i) $T \nvdash \xi(k)$,
(ii) $T \vdash \xi(k+1) \to \xi(k)$,
(iii) $\xi(k)$ is Σ_n-conservative over $T + \neg\xi(k+1)$.

Proof. In a first attempt to prove this it is natural to let $\xi(x)$ be such that

$$PA \vdash \xi(k) \leftrightarrow \xi(k+1) \vee \forall v([\Sigma_n](\neg\xi(k+1)\wedge\xi(k),v) \rightarrow \neg Prf_T(\xi(k),v)).$$

But then (i) does not follow and so we have to proceed in a more indirect way.

Let $\delta(u)$ be any formula. Let $\kappa(z,x,y)$ be a Π_n formula such that
(1) $\quad PA \vdash \neg\kappa(z,x,0)$,
(2) $\quad PA \vdash \kappa(\delta,k,y+1) \leftrightarrow \kappa(\delta,k+1,y) \vee$
$$\forall v([\Sigma_n](\neg\eta_\delta(k)\wedge\xi_\delta(k),v) \rightarrow \neg Prf_T(\xi_\delta(k),v)),$$
where
$$\xi_\delta(x) := \forall u(\delta(u) \rightarrow \kappa(\delta,x,(u \dotminus x) + 1)),$$
$$\eta_\delta(x) := \forall u(\delta(u) \rightarrow \kappa(\delta,x+1,u \dotminus x)).$$
($\dotminus$ is the function such that $k \dotminus m = k - m$ if $k \geq m$ and $= 0$ otherwise.) In (2) set $y = u \dotminus k$. Then, since neither y nor u is free in the second disjunct of (2), by predicate logic, we get
(3) $\quad PA \vdash \xi_\delta(k) \leftrightarrow \eta_\delta(k) \vee \forall v([\Sigma_n](\neg\eta_\delta(k)\wedge\xi_\delta(k),v) \rightarrow \neg Prf_T(\xi_\delta(k),v)).$
It follows that
(4) $\quad$ if $T \vdash \xi_\delta(k)$, then $T \vdash \eta_\delta(k)$.
For let p be a proof of $\xi_\delta(k)$ in T. By Lemma 1(ii),
$$T + \neg\eta_\delta(k) \wedge \xi_\delta(k) \vdash \neg Prf_T(\xi_\delta(k),p),$$
whence $T + \xi_\delta(k) \vdash \eta_\delta(k)$ and so $T \vdash \eta_\delta(k)$.

Clearly
(5) $\quad$ if $T \vdash \delta(u) \rightarrow u > k$, then $T \vdash \eta_\delta(k) \leftrightarrow \xi_\delta(k+1)$.
Suppose now $\delta(u)$ is PR. Then
(6) $\quad$ if $\exists u \delta(u)$ is true, then $T \nvdash \xi_\delta(0)$.
Suppose $\exists u \delta(u)$ is true and $T \vdash \xi_\delta(0)$. Let m be the least number such that $\delta(m)$ is true. Then $T \vdash \delta(u) \rightarrow u \geq m$. By (4) and (5), it follows that $T \vdash \eta_\delta(m)$. But also $T \vdash \delta(m)$ and so, by (1), $T \vdash \neg\eta_\delta(m)$, a contradiction. Thus, (6) is proved.

Now let $\delta'(u)$ be a PR formula such that
(7) $\quad PA \vdash \exists u \delta'(u) \leftrightarrow Pr_T(\xi_{\delta'}(0))$.
If $\exists u \delta'(u)$ is true, then, by (6), $Pr_T(\xi_{\delta'}(0))$ is false and, by (7), it is true. Thus, $\exists u \delta'(u)$ is false, whence, by (7), $Pr_T(\xi_{\delta'}(0))$ is false and so $T \nvdash \xi_{\delta'}(0)$.

Let $\xi(x) := \xi_{\delta'}(x)$ and $\eta(x) := \eta_{\delta'}(x)$. Then $T \nvdash \xi(0)$. Hence, by (3) and (5) with $\delta(u) := \delta'(u)$, we get (i) and (ii).

(iii) can be verified as follows. Suppose
(8) $\quad T + \neg\xi(k+1) + \xi(k) \vdash \sigma$,
where σ is Σ_n. Then, by (5), $T + \neg\eta(k) + \xi(k) \vdash \sigma$. Hence, by Lemma 1(iii), there is a q such that
$$T + [\Sigma_n](\neg\eta(k)\wedge\xi(k),q) \vdash \sigma.$$

But then, by (i), (3), and Lemma 1(i), $T + \neg\sigma \vdash \xi(k)$, whence $T + \neg\xi(k) \vdash \sigma$ and so, by (8), $T + \neg\xi(k+1) \vdash \sigma$, proving (iii). ∎

Proof of Theorem 4.8. Let $\xi(x)$ be as in Lemma 8. By Lemma 8(i) and (iii), $T \nvdash \xi(k) \to \xi(k+1)$. It follows that $T + \xi(0) + \{\xi(k) \to \xi(k+1): k \in N\}$ is an axiomatization of $T + \{\xi(k): k \in N\}$ which is irredundant over T. Let π_k, $k \in N$, be Π_n sentences such that $T + \{\pi_k: k \in N\}$ is an axiomatization of $T + \{\xi(k): k \in N\}$. Let r be arbitrary. By Lemma 8(ii), there is an m such that $T + \xi(m) \vdash \pi_r$. Let s be such that $T + \pi_0 \wedge...\wedge \pi_s \vdash \xi(m+1)$. We may assume that $s > r$. It follows that

$$T + \xi(m) \wedge \neg\xi(m+1) \vdash \neg(\pi_0 \wedge...\wedge \pi_{r-1} \wedge \pi_{r+1} \wedge...\wedge \pi_s).$$

But then, by Lemma 8(iii),

$$T + \pi_0 \wedge...\wedge \pi_{r-1} \wedge \pi_{r+1} \wedge...\wedge \pi_s \vdash \xi(m+1).$$

It follows, by Lemma 8(ii), that $T + \{\pi_k: k \neq r\} \vdash \pi_r$. Thus, $T + \{\pi_k: k \in N\}$ is not irredundant over T. ∎

We have actually proved more than is stated in Theorem 4.8. First of all, for every r, $T + \{\pi_k: k \neq r\} \vdash \pi_r$. Secondly, this holds for all, not necessarily r.e., sets $\{\pi_k: k \in N\}$ of Π_n sentences such that $T + \{\pi_k: k \in N\} \dashv\vdash T + \{\xi(k): k \in N\}$. The theory $T + \{\eta(k): k \in N\}$ constructed in the proof of Theorem 4.7, on the other hand, is deductively equivalent to $T + \{\eta(k): k \notin H\}$ and $\{\eta(k): k \notin H\}$ is irredundant over T. (The set $\{\eta(k): k \notin H\}$ is not r.e. (cf. Lemma 4.6).)

§2. Partial conservativity and propositional logic.

In this § we prove some general results on Π_n-conservativity, from which some of the results proved in §1 can be obtained as special cases. This will also give us the opportunity to give an example of a very fruitful and versatile technique, that of defining a suitable so-called Solovay function and then defining, using this function, an arithmetical interpretation (of propositional formulas).

Let PF_r be the set of formulas of propositional logic in the variables $p_0,...,p_r$. We use F, G, etc. for members of PF_r. An *arithmetical interpretation* I assigns to each variable p_i an arithmetical sentence p_i^I and commutes with the propositional connectives, i.e.,

$$(\neg F)^I = \neg F^I, (F \wedge G)^I = F^I \wedge G^I, \text{etc.}, \bot^I := \bot.$$

In what follows we write $\varphi \dashv_T \psi$ for $T + \varphi \dashv_T T + \psi$.

Given a set X of ordered pairs (F,G) we may ask under what conditions there is an interpretation I such that for all F, G, $F^I \dashv_{\Pi_n} G^I$ iff $(F,G) \in X$. This question is answered in Theorem 7 and Exercise 13, below.

By a *model for Π-conservativity* (*for PF_r*) we understand a triple **M** =

$(V, \leq, \Vdash)$, where V is a finite nonempty set of natural numbers, $0 \in V$, $\leq$ is a reflexive and transitive relation on V, $0 \leq a$ for every $a \in V$, and $\Vdash$ is an arbitrary relation between members of V and propositional variables p_i, $i \leq r$. We use a, b, etc. to denote members of V. Members of V can be thought of as truth-value assignments: a assigns the value true to p_i iff $a \Vdash p_i$. $a \Vdash F$, where F is any formula, is then defined by:

$a \Vdash \neg F$ iff $a \nVdash F$, $a \Vdash F \wedge G$ iff $a \Vdash F$ and $a \Vdash G$, etc.

We write $\Vdash (F,G)$ to mean that for every $a \in V$, if $a \Vdash G$, there is a $b \in V$ such that $a \leq b$ and $b \Vdash F$.

In what follows we sometimes say that a sentence φ is Γ if, although this is not literally true, φ is provably in PA equivalent to a Γ sentence.

Theorem 7. For every model **M** for Π-conservativity, there is an interpretation I such that for all F, G, $F^I \dashv_{\Pi_n} G^I$ iff $\Vdash (F,G)$.

The relation $\dashv_\Gamma$ can be characterized as follows (compare Lemma 6.2). Note that $T \vdash \varphi$ iff $\bot \dashv_\Gamma \neg \varphi$.

Lemma 9. $\varphi \dashv_\Gamma \psi$ iff for every k, $T + \psi \vdash [\Gamma](\varphi,k)$.

Proof. Suppose first $\varphi \dashv_\Gamma \psi$. By Lemma 1(ii), for every k, $T + \varphi \vdash [\Gamma](\varphi,k)$. Now, $[\Gamma](\varphi,k)$ is Γ. Hence, by assumption, $T + \psi \vdash [\Gamma](\varphi,k)$, as desired.

To prove the converse implication, let θ be Γ sentence such that $T + \varphi \vdash \theta$. By Lemma 1(iii), there is then a k such that $T + [\Gamma](\varphi,k) \vdash \theta$. By assumption, $T + \psi \vdash [\Gamma](\varphi,k)$ and so $T + \psi \vdash \theta$, as desired. ∎

The main idea of the proof of Theorem 7 is to define in PA (and in the real world) a certain (Solovay) function f: $N \to V$ and then define I in the following way. Let

$\lambda(x) := \exists y \forall z \geq y (f(z) = x)$.

Thus, $\lambda(x)$ "says" that x is the limit of f(z) as $z \to \infty$. Now let

$p_i^I := \bigvee \{\lambda(a): a \Vdash p_i\}$.

(The disjunction of the empty set is defined to be $\bot$.)

There is a Σ_{n-1} formula $\delta(u,w)$ such that $PA \vdash Tr_{\Pi_n}(u) \leftrightarrow \forall w \delta(u,w)$. We shall assume (in the proof of Lemma 12(vii)) that $PA \vdash \neg \delta(\bot,w)$: if necessary, replace $\delta(u,w)$ by $\delta(u,w) \wedge u \neq \bot$. Let

rank$_n$(a,s) = the least $y \leq s$ for which there is a $w \leq s$ such that

$\exists u \leq y (\Pi_n(u) \wedge Prf_{T+\lambda(a)}(u,v) \wedge \neg \delta(u,w))$, if y exists,

= ∞ otherwise.

Here ∞ is a fictitious "number" > all natural numbers. We leave it to the

§2. Partial Conservativity and Propositional Logic

reader to get rid of ∞. In what follows we usually write rank(a,s) for $\text{rank}_n(a,s)$.

Using self-reference we can now define in PA a function f satisfying the following conditions (compare the proof of Theorem 3.5). In the definition of f in PA the relation $\preccurlyeq$ is represented by the formula

$$x \preccurlyeq y := \bigvee \{x = a \wedge y = b : a \preccurlyeq b\}.$$

$\preccurlyeq$ is then reflexive and transitive provably in PA.

Stage 0. $f(0) = 0$.
Stage s+1. Suppose $f(s) = a$.
   Case 1. There is a (least) b such that $a \preccurlyeq b$ and $\text{rank}(b,s) < \text{rank}(a,s)$.
      $f(s+1) = b$.
   Case 2. Otherwise.
      $f(s+1) = a$.

The occurrence of "least" in Case 1 is there just to ensure that $f(s+1)$ is uniquely defined. Note that f does not depend on $\Vdash$.

The main lemmas, from which Theorem 7 follows quite easily, are:

Lemma 10. (a) $\text{PA} \vdash \bigvee \{\lambda(a) : a \in V\}$.
  (b) For all $a, b \in V$, if $a \neq b$, then $\text{PA} \vdash \neg(\lambda(a) \wedge \lambda(b))$.

Lemma 11. Suppose $a, b \in V$.
  (a) If $a \preccurlyeq b$, then $\lambda(b) \dashv_{\Pi_n} \lambda(a)$.
  (b) $\neg \bigvee \{\lambda(c) : a \preccurlyeq c\} \dashv_{\Pi_n} \lambda(a)$.

The definition of f can be explained, at least in part, as follows. From $a \preccurlyeq b$, we want to be able to conclude that $\lambda(b) \dashv_{\Pi_n} \lambda(a)$ (Lemma 11(a)). The *prima facie* simplest way to ensure this would be to allow f to move from a (= $f(s)$) to b (= $f(s+1)$) provided only that
(+) $\exists uvw \leqslant s(\Pi_n(u) \wedge \text{Prf}_{T+\lambda(b)}(u,v) \wedge \neg\delta(u,w))$
(and b is minimal). Then we would have $T + \lambda(a) \vdash \forall v[\Pi_n](\lambda(b),v)$ and so, by Lemma 9, $\lambda(b) \dashv_{\Pi_n} \lambda(a)$ (and the formula $f(s) = y$ would be Σ_n; see Lemma 14, below). But we would then not be able to prove Lemma 10(a). Thus, in Case 1 (of the definition of f) we need a requirement which is stronger than (+) but still does not block the proof of Lemma 11(a).

Lemma 12. For all k and all $a \in V$,
  (i) $\text{PA} \vdash x \leqslant y \rightarrow \text{rank}(a,y) \leqslant \text{rank}(a,x)$,
  (ii) $\text{PA} \vdash \text{rank}(a,x) \leqslant x \vee \text{rank}(a,x) = \infty$,
  (iii) $\text{PA} \vdash x \leqslant y \rightarrow \text{rank}(f(y),y) \leqslant \text{rank}(f(x),x)$,

(iv) $PA \vdash [\Pi_n](\lambda(a),x) \to \mathrm{rank}(a,y) > x$,
(v) $T \vdash \lambda(a) \to \mathrm{rank}(a,x) > k$,
(vi) $PA \vdash \neg[\Pi_n](\lambda(a),x) \to \exists y(\mathrm{rank}(a,y) \leq x)$,
(vii) $PA \vdash \mathrm{Prf}_{T+\lambda(a)}(\bot,x) \to \mathrm{rank}(a,x) \leq x$,
(viii) $T \vdash \mathrm{rank}(0,k) > k$.

Proof. (i) and (ii) are immediate. ♦

(iii) It is easily checked that $PA \vdash \mathrm{rank}(f(x+1),x+1) \leq \mathrm{rank}(f(x),x)$. Now use induction. ♦

(iv) This is clear. ♦

(v) This follows from (iv) and Lemma 1(ii). ♦

(vi) Reason in PA: "Suppose $\neg[\Pi_n](\lambda(a),x)$. Let w be such that
$$\exists uv \leq z(\Pi_n(u) \wedge \mathrm{Prf}_{T+\lambda(a)}(u,v) \wedge \neg\delta(u,w)).$$
Let $y = \max\{x,w\}$. Then $\mathrm{rank}(a,y) \leq x$." ♦

(vii) Reason in PA: "Suppose $\mathrm{Prf}_{T+\lambda(a)}(\bot,x)$. We have assumed that $\neg\delta(\bot,w)$ and so, in particular, $\neg\delta(\bot,0)$. But then
$$\Pi_n(\bot) \wedge \mathrm{Prf}_{T+\lambda(a)}(\bot,x) \wedge \neg\delta(\bot,0)),$$
whence, since $\bot \leq x$, $\mathrm{rank}(a,x) \leq x$." ♦

(viii) Reason in T: "By Lemma 10(a), $\lambda(a)$ for some a. If $a = 0$, then, by (v), $\mathrm{rank}(0,k) > k$. Suppose $a \neq 0$. Let x be such that $f(x) = 0$ and $f(x+1) \neq 0$. Let $b = f(x+1)$. Since $\lambda(a)$, there is a $y \geq x+1$ such that $f(y) = a$. Thus, by (v), (iii), (i), $k < \mathrm{rank}(a,y) \leq \mathrm{rank}(b,x+1) \leq \mathrm{rank}(b,x)$. Case 1 applies at $x+1$, whence $\mathrm{rank}(b,x) < \mathrm{rank}(0,x)$, whence $\mathrm{rank}(b,x) < \infty$ and so, by (ii), $\mathrm{rank}(b,x) \leq x$. It follows that $k \leq x$ and so, by (i), $\mathrm{rank}(0,x) \leq \mathrm{rank}(0,k)$. Combining these inequalities we get $k < \mathrm{rank}(0,k)$, as desired." (It follows that $T \vdash \mathrm{rank}(0,k) = \infty$. Also, since in this proof "$k \leq x$", for every n, $T \vdash f(n) = 0$.) ∎

Note that from Lemma 12(iv) and (vi) it follows that
$$PA \vdash [\Pi_n](\lambda(a),x) \leftrightarrow \forall y(\mathrm{rank}(a,y) > x).$$

Proof of Lemma 10. (a) Argue in PA: "Lemma 12(iii), $\mathrm{rank}(f(x),x)$ is non-increasing. Also, every time Case 1 applies $\mathrm{rank}(f(x),x)$ decreases. It follows that $f(x)$ cannot change value infinitely often." ♦

(b) This is obvious. ∎

Lemma 13. Suppose $a, b \in V$.
(a) If $a \leq b$, then for every k, $T + \lambda(a) \vdash [\Pi_n](\lambda(b),k)$.
(b) $PA \vdash f(x) = a \to \bigvee\{\lambda(c): a \leq c\}$,
(c) $T + \lambda(a)$ is consistent.

§2. Partial Conservativity and Propositional Logic

Proof. (a) Argue in $T + \lambda(a)$: "Suppose $\neg[\Pi_n](\lambda(b),k)$. Then, by Lemma 12(vi), (i), there is a y such that for $x \geqslant y$, $\text{rank}(b,x) \leqslant k$. There is a $z \geqslant y$ such that $f(x) = a$ for $x \geqslant z$. Also, by Lemma 12(v), $\text{rank}(a,z) > k$. It follows that $\text{rank}(b,z) < \text{rank}(a,z)$. But then Case 1 applies at Stage $z+1$ and so $f(z+1) \neq a$ (although not necessarily $f(z+1) = b$), a contradiction." ♦

(b) $\preccurlyeq$ is reflexive and transitive provably in PA. Clearly, $\text{PA} \vdash f(x) \preccurlyeq f(x+1)$. By induction $\text{PA} \vdash x \leqslant y \rightarrow f(x) \preccurlyeq f(y)$. It follows that $\text{PA} \vdash f(x) = a \wedge \lambda(b) \rightarrow a \preccurlyeq b$. Now use Lemma 10(a). ♦

(c) Let k be any natural number. By Lemma 12(vii),
$$\text{PA} \vdash \text{Prf}_{T+\lambda(0)}(\bot,k) \rightarrow \text{rank}(0,k) \leqslant k.$$
But then, by Lemma 12(viii), $T \vdash \neg\text{Prf}_{T+\lambda(0)}(\bot,k)$. Since this is true for every k, it follows that $T + \lambda(0)$ is consistent.

Next, suppose $a \neq 0$. Since $T + \lambda(0)$ is consistent, it suffices to show that for every k,
$$T + \lambda(0) \vdash \neg\text{Prf}_{T+\lambda(a)}(\bot,k).$$
Reason in $T + \lambda(0)$: "$0 \preccurlyeq a$. Hence, by (a), $[\Pi_n](\lambda(a),k)$ and so, by Lemma 12(iv), $\text{rank}(a,k) > k$, whence, by Lemma 12(vii), $\neg\text{Prf}_{T+\lambda(a)}(\bot,k)$." ∎

Lemma 14. The formula $f(x) = y$ is Δ_n.

Proof. The formula $\delta(u,w)$ is Σ_{n-1}. Hence, by Lemma 12(ii), the formula $\text{rank}(x,z) < \text{rank}(y,z)$ is Δ_n. It follows that the formula $f(x) = y$ is Σ_n (cf. Chapter 1, p. 11). But then, since f is total provably in PA, $f(x) = y$ is Δ_n. ∎

Proof of Lemma 11. (a) By Lemma 9, it suffices to show that for every k,
$$T + \lambda(a) \vdash [\Pi_n](\lambda(b),k).$$
But this is just Lemma 13(a). ♦

(b) By Lemma 14, $\pi := \forall x(f(x) \neq a)$ is a Π_n sentence. From Lemma 13(b) it follows that
$$\text{PA} + \neg\bigvee\{\lambda(b): a \preccurlyeq b\} \vdash \pi.$$
Clearly, $T + \lambda(a) \vdash \neg\pi$ and so, by Lemma 13(c), $T + \lambda(a) \nvdash \pi$. ∎

Proof of Theorem 7. Let I be defined by:
$$p_i^I := \bigvee\{\lambda(a): a \Vdash p_i\}.$$
Then for every E,
(1) if $a \Vdash E$, then $\text{PA} \vdash \lambda(a) \rightarrow E^I$,
(2) if $a \nVdash E$, then $\text{PA} \vdash \lambda(a) \rightarrow \neg E^I$.
For $E := p_i$ this follows from Lemma 10(b). The inductive steps are straightforward.

From (1), (2), and Lemma 10(a) it follows that
(3) $\text{PA} \vdash E^I \leftrightarrow \bigvee\{\lambda(a): a \Vdash E\}$.

(This includes the case where the disjunction is empty and so $PA \vdash \neg E^I$.)

Now suppose $\Vdash (F,G)$. We want to show that $F^I \dashv_{\Pi_n} G^I$. Let π be a Π_n sentence such that $T + F^I \vdash \pi$. Suppose a $\Vdash G$. By (3) with $E := G$, it is then sufficient to show that $T + \lambda(a) \vdash \pi$. (This is sufficient even if there is no a for which a $\Vdash G$, since then $PA \vdash \neg G^I$ and so $F^I \dashv_{\Pi_n} G^I$ holds trivially.) Now, since $\Vdash (F,G)$, there is a b such that $a \preccurlyeq b \Vdash F$. By (3) with $E := F$, $T + \lambda(b) \vdash F^I$ and so $T + \lambda(b) \vdash \pi$. But then, by Lemma 11(a), $T + \lambda(a) \vdash \pi$, as desired.

Now suppose $\nVdash (F,G)$. Let a be such that a $\Vdash G$ and for every b, if $a \preccurlyeq b$, then $b \nVdash F$. Then, by (3),
$$PA \vdash \lambda(a) \to G^I,$$
$$PA \vdash F^I \to \neg \bigvee \{\lambda(c): a \preccurlyeq c\}.$$
It follows, by Lemma 11(b), that $F \nvdash_{\Pi_n} G^I$, as desired. ∎

As an example of an application of Theorem 7 we prove the following:

Corollary 2. Let F_i, $i \leq p$, and G_j, $j \leq q$, be propositional formulas such that no formula $F_i \to G_j$ is a tautology. There is then an interpretation I such that $F_i^I \dashv_{\Pi_n} T$ for $i \leq p$, and $G_j^I \nvdash_{\Pi_n} T$ for $j \leq q$.

Proof. Let $t_{i,j}$ be a truth-value assignment (in the usual sense) to the variables p_m, $m \leq r$, such that $t_{i,j}(F_i)$ = true and $t_{i,j}(G_j)$ = false. Let $V = \{0\} \cup \{\langle i,j \rangle : i \leq p \ \& \ j \leq q\}$. Let $\langle i,j \rangle \preccurlyeq \langle i',j' \rangle$ iff $j = j'$. Then $\preccurlyeq$ is reflexive and transitive. Also, $0 \preccurlyeq a$ for every $a \in V$. Let $\langle i,j \rangle \Vdash p_m$ iff $t_{i,j}(p_m)$ = true. Then for every F, $\langle i,j \rangle \Vdash F$ iff $t_{i,j}(F)$ = true. It follows that $\langle i,j \rangle \Vdash F_i \wedge \neg G_j$. If $\langle i',j' \rangle \in V$, then $\langle i',j' \rangle \preccurlyeq \langle i,j' \rangle \Vdash F_i$. It follows that, regardless of how $0 \Vdash p_i$ is defined, $\Vdash (F_i, T)$. Also $\langle i,j \rangle \nVdash G_j$ for all $i \leq p$ and so $\nVdash (G_j, T)$. Now use Theorem 7. ∎

Let us say that an arithmetical interpretation I is Φ if the sentences p_i^I are Φ. Let I be as in the proof of Theorem 7. By Lemma 14, $\lambda(x)$ is Σ_{n+1}. Also, by Lemma 10(a),
$$PA \vdash \lambda(x) \leftrightarrow \forall y \exists z \geq y (f(z) = x).$$
Thus, the formula $\lambda(x)$ is Δ_{n+1} and so I is Δ_{n+1}. For some applications, however, this is not good enough: we want I to be Σ_n. Such interpretations are provided by the following result.

The relation $\preccurlyeq$ (on V) is said to be *truth-preserving (with respect to $\Vdash$)* if for all a, b $\in V$ and all p_i, if a $\Vdash p_i$ and $a \preccurlyeq b$, then b $\Vdash p_i$.

Theorem 8. Let M be any model for Π-conservativity such that $\preccurlyeq$ is truth-preserving. There is then a Σ_n interpretation I such that for all F, G,

§2. Partial Conservativity and Propositional Logic

$F^I \dashv_{\Pi_n} G^I$ iff $\Vdash (F, G)$.

Proof. Let f and I be defined in the same way as in the proof of Theorem 7. Since $\preccurlyeq$ is truth-preserving, by Lemma 10(c),

$$PA \vdash p_i^I \leftrightarrow \bigvee \{\exists x (f(x) = a) : a \Vdash p_i\}.$$

Thus, I is Σ_n. ∎

As an application of Theorem 8 we now prove the following corollary (which can also be derived from Theorems 2(a) and 5(a)).

Corollary 3. There are Σ_n sentences σ_0, σ_1 such that
(i) $T \nvdash \sigma_0 \vee \sigma_1$,
(ii) σ_i is Π_n-conservative over $T + \neg\sigma_{1-i}$, $i = 0, 1$,
(iii) $\sigma_0 \wedge \sigma_1$ is Π_n-conservative over T.

Proof. Let V = {0,1,2,3}. Let a $\preccurlyeq$ b iff (i) a = b or (ii) a = 0 and either b = 1 or b = 2 or (iii) (a = 1 or a = 2) and b = 3. Define $\Vdash$ by: 0 $\Vdash p_i$ for i = 0, 1, 1 $\Vdash p_i$ iff i = 0, 2 $\Vdash p_i$ iff i = 1, and 3 $\Vdash p_i$ for i = 0, 1. Then $(V, \preccurlyeq, \Vdash)$ is as assumed in Theorem 8. Also $\Vdash (\bot, \neg p_0 \wedge \neg p_1)$, $\Vdash (p_0 \wedge \neg p_1, \neg p_1)$, $\Vdash (\neg p_0 \wedge p_1, \neg p_0)$, and $\Vdash (p_0 \wedge p_1, T)$. Now, let I be a Σ_n interpretation such that $F^I \dashv_{\Pi_n} G^I$ iff $\Vdash (F, G)$. Let $\sigma_i := p_i^I$, i = 0, 1. Then conditions (i), (ii), (iii) are satisfied. ∎

We conclude by giving a proof of Theorem 3(a) using a method which has many other applications (but is far from complete). In the proof we use the obvious fact that $\varphi \dashv_\Gamma \psi$ iff $\varphi \dashv_\Gamma \neg\varphi \wedge \psi$; in particular, $\varphi \dashv_\Gamma T$ iff $\varphi \dashv_\Gamma \neg\varphi$.

Let $\beta(u,w)$ be a Π_{n-1} formula such that $PA \vdash Tr_{\Sigma_n}(u) \leftrightarrow \exists w \beta(u,w)$. Let $\{\Sigma_n\}(x,u,v,w) :=$

$$\Sigma_n(u) \wedge Prf_{T+x}(u,v) \rightarrow \beta(u,w).$$

Then

$$PA \vdash [\Sigma_n](x,y) \leftrightarrow \forall uv \leqslant y \exists w \{\Sigma_n\}(x,u,v,w).$$

Alternative proof of Theorem 3(a). Let f be defined in such a way that, provably in PA:

Stage 0. f(0) = 0.
Stage s+1. Case 1. f(s) = 0, $rank_n(1,s) \leqslant s$, and

$$\forall uv \leqslant rank_n(1,s) \exists w \leqslant s \{\Sigma_n\}(\lambda(0),u,v,w).$$

    f(s+1) = 1.
Case 2. Otherwise.
    f(s+1) = f(s).
Clearly,

$$PA \vdash \lambda(0) \leftrightarrow \neg\lambda(1).$$

Let $\varphi := \lambda(1)$. Then $PA \vdash \varphi \leftrightarrow \exists x(f(x) = 1)$ and so φ is Σ_n. By Lemma 9, to prove that φ is Π_n-conservative over T, it is sufficient to show that for every k, $T + \lambda(0) \vdash [\Pi_n](\lambda(1),k)$. Argue in $T + \lambda(0)$: "Since $\lambda(0)$, $f(x) = 0$ for all x. Suppose $\neg[\Pi_n](\lambda(1),k)$. Then $\text{rank}_n(1,x) \leq k$ for sufficiently large x. Since $\lambda(0)$, we have $[\Sigma_n](\lambda(0),k)$ and so $\forall uv \leq k \exists w \leq x\{\Sigma_n\}(\lambda(0),u,v,w)$ for sufficiently large x. It follows that Case 1 applies at some Stage x+1. But then $f(x+1) = 1$, a contradiction."

$PA \vdash \neg\varphi \leftrightarrow \lambda(0)$. Thus, to prove that $\neg\varphi$ is Σ_n-conservative over T, it is sufficient to show that for every k, $T + \lambda(1) \vdash [\Sigma_n](\lambda(0),k)$. Argue in $T + \lambda(1)$: "There is an x such that $f(x) = 0$ and $f(x+1) = 1$. Case 1 applies at x+1. Also, $\text{rank}_n(1,x) \geq k$. It follows that $[\Sigma_n](\lambda(0),k)$." ∎

Exercises for Chapter 5
In the following Exercises we assume that $PA \dashv T$.

1. Let θ be a Π_1 Rosser sentence for T. Show that $\neg\theta$ is not Π_1-conservative over T. [Hint: Use Exercise 2.12(b).]

2. Suppose T is not Σ_1-sound.
   (a) Show that Con_T is not Σ_1-conservative over T. [Hint: Let $\delta(y)$ be PR and such that $\exists y \delta(y)$ is false and provable in T. Let χ be such that
   $$T \vdash \chi \leftrightarrow \exists y(\delta(y) \land \forall z \leq y \neg \text{Prf}_T(\chi,z)).$$
   $T \nvdash \chi$. Let $\chi^* := \exists z(\text{Prf}_T(\chi,z) \land \forall y \leq z \neg \delta(y))$. Then $T \vdash \chi^* \leftrightarrow \neg\chi$. It follows that $T \vdash \neg\chi \to \text{Pr}(\chi) \land \text{Pr}(\neg\chi)$.]
   (b) Improve (a) by showing that if θ is a Π_1 Rosser sentence for T, θ is not Σ_1-conservative over T. [Hint: Let $\psi := \exists u(\text{Prf}_T(\neg\theta,u) \land \forall z \leq u \neg \text{Prf}_T(\theta,z))$. $T + \neg\psi$ is consistent. $T + \neg\psi + \theta \vdash \text{Con}_{T+\neg\theta}$ and $T + \neg\theta \vdash \neg\psi$. Thus, $T + \neg\psi + \theta \vdash \text{Con}_{T+\neg\psi}$. Apply (a) to $T + \neg\psi$.]

3. Show that the result of replacing Σ_n by Π_n in Corollary 1 is false.

4. φ is a *self-prover in* T if $T \vdash \varphi \to \text{Pr}_T(\varphi)$. Every Σ_1 sentence is a self-prover.
   (a) Show that φ is a self-prover in T iff there is a sentence θ such that $T \vdash \varphi \leftrightarrow (\theta \land \text{Pr}_T(\theta))$.
   (b) Show that if $\Gamma \neq \Pi_1$, there is a Γ self-prover in T which is not $\Gamma^{d,T}$. [Hint: Let θ (see (a)) be Γ^d-conservative over T; use Löb's theorem.]

5. (a) Show that clause (ii) of Lemma 2 can be replaced by

if PA ⊣ S ⊣ T, then S + ξ(k) ⊣_Γ dS + {χ(k,q): q ∈ N}.

(b) φ is *hereditarily* Γ-*conservative* over T if φ is Γ-conservative over S for every S such that PA ⊣ S ⊣ T. Show that in Lemma 3 and Theorem 2 we can replace "Γ-conservative" by "hereditarily Γ-conservative".

(c) Show that in Theorem 3 we cannot in general replace "Γ ($Γ^d$)-conservative" by "hereditarily Γ ($Γ^d$)-conservative". [Hint: Let φ be a $Σ_1$ sentence and ψ a $Π_1$ sentence such that PA + φ ∧ ψ is consistent and PA ⊬ φ ∨ ψ. Let T = PA + φ ∧ ψ.]

6. (a) Show that there is a $Δ_{n+1}$ sentence φ such that φ and ¬φ are $Π_n$-conservative over T. [Hint: Let φ be such that
$$PA ⊢ φ ↔ ∃y(¬[Π_n](φ,y) ∧ [Π_n](¬φ,y)).]$$

(b) Show that if T is $Σ_n$-sound, there is no $Δ_{n+1}$ sentence φ such that φ and ¬φ are $Σ_n$-conservative over T. [Hint: If T is $Σ_n$-sound, φ is $Π_{n+1}$, and φ is $Σ_n$-conservative over T, then φ is true.]

(c) Show that there is no B_n sentence φ such that φ and ¬φ are $Π_n$- ($Σ_n$-) conservative over T. Conclude that there is a $Δ_{n+1}$ sentence which is not B_n^T (see Corollary 2.6). [Hint: Suppose not. Let φ := $(π_0 ∧ σ_0) ∨ ... ∨ (π_n ∧ σ_n)$. In the $Π_n$ case, for k ≤ n+1, show that
$$T ⊢ ∨\{\bigwedge_{j∈X} ¬σ_j : X ⊆ \{0,...,n\} \text{ \& } X \text{ has exactly k elements}\}.]$$

7. Let X_0 and X_1 be disjoint r.e. sets.

(a) Show that there is a $Σ_n$ formula ξ(x), such that $ξ^i(x)$ numerates X_i in T, i = 0, 1, and if k ∉ $X_0 ∪ X_1$, then ξ(k) is $Π_n$-conservative over T and ¬ξ(k) is $Σ_n$-conservative over T.

(b) Show that there is a formula η(x) such that all the sentences (¬)η(0) ∧...∧ (¬)η(k) are Γ-conservative over T.

(c) Show that there is a formula ξ(x) such that (i) if k ∈ X_0, then T ⊢ ξ(k), (ii) if k ∈ X_1, then T ⊢ ¬ξ(k), (iii) if Y_0 and Y_1 are any disjoint finite subsets of $(X_0 ∪ X_1)^c$, then ∧{ξ(k): k ∈ Y_0} ∧ ∧{¬ξ(k): k ∈ Y_1} is Γ-conservative over T. [Hint: Let η(x) be as in (b). Next let $ξ_i(x)$ be as in Lemma 3 with Γ replaced by $Γ^d$ and X = X_i. Let $ρ_i(x,y)$ be a PR formula such that X_i = {k: ∃m Q ⊢ $ρ_i(k,m)$}, i = 0, 1. Let
$$γ(x) := ∃y(ρ_0(x,y) ∧ ∀z≤y ¬ρ_1(x,z)).$$
Finally, let ξ(x) := $(¬ξ_0(x) ∨ η(x)) ∧ (ξ_1(x) ∨ γ(x))$.]

8. (a) Let X and Y be r.e. sets of Γ and $Γ^d$ sentences, respectively, such that if φ ∈ X and ψ ∈ Y, then T ⊢ φ ∨ ψ. Show that there is a Γ sentence θ such that T + θ is a $Γ^d$-conservative extension of T + X and T + ¬θ is a Γ-conserva-

tive extension of $T + Y$.

(b) Let $\theta_0, \theta_1, \theta_2, \ldots$ be a recursive sequence of Γ sentences such that $T \nvdash \neg \theta_k$ and $T \vdash \neg(\theta_k \wedge \theta_m)$ for $k \neq m$. Let X_0 and X_1 be disjoint r.e. sets. Show that there is a sentence φ such that $X_0 = \{k: T \vdash \theta_k \to \varphi\}$ and $X_1 = \{k: T \vdash \theta_k \to \neg\varphi\}$.

9. (a) Suppose φ is Σ_n, and Π_n-conservative over T. Let ψ be any Π_n sentence which is Σ_n-conservative over $T + \varphi$. Show that $T + \neg\varphi \vdash \psi$. Conclude that no Π_n sentence is nontrivially Σ_n-conservative over $T + \varphi$ and $T + \neg\varphi$. [Hint: Let $\varphi := \exists x \gamma(x)$ and $\psi := \forall x \delta(x)$, where $\gamma(x)$ and $\delta(x)$ are Π_{n-1} and Σ_{n-1}, respectively. Then $T + \varphi + \psi \vdash \exists x(\gamma(x) \wedge \forall y \leqslant x \delta(y))$.]

(b) Show that there is an r.e. family of consistent extensions of PA such that for no Γ does there exist a Γ sentence which is nontrivially Γ^d-conservative over every member of the family. [Hint: Let φ be a Π_1 sentence undecidable in PA. Then
$$\{PA + \neg\theta: PA \vdash \theta \to \varphi\} \cup \{PA + \theta: PA \vdash \varphi \to \theta\}$$
is an r.e. family of extensions of PA. Suppose θ is Π_n and nontrivially Σ_n-conservative over all members of this family. Then $PA + \varphi \nvdash \theta$. θ is Σ_n-conservative over $PA + \neg(\theta \wedge \varphi)$. It follows that $PA + \varphi \vdash \theta$, a contradiction. The dual case is similar.]

10. This exercise may be compared with Theorems 2.13, 2.14.

(a) For each Γ, there is a primitive recursive function f such that for every Γ sentence φ, $f(\varphi)$ is a proof in PA of $\varphi \leftrightarrow Tr_\Gamma(\varphi)$. Use this to show that there are a Γ sentence θ and a primitive recursive function g such that $T \nvdash \theta$ and if ψ is any Γ^d sentence and q a proof of ψ in $T + \theta$, then $g(q)$ is a proof of ψ in T.

(b) Let f be any recursive function. Show that there are sentences φ, ψ such that φ is Γ-conservative over T, ψ is Γ, $T \vdash \psi$, and there is a proof p of ψ in $T + \varphi$ such that $q > f(p)$ for every proof q of ψ in T.

11. Show that there is a theory $S \vdash T$ which is e.u. over T but not effectively e.i. over T. (Compare Exercise 4.8.) [Hint: Let H be as in Lemma 4.6. Let $\xi_n(x)$ be the formula $\xi(x)$ of Lemma 3 with $X = H$ and $\Gamma = \Sigma_n$. Let $S = T + \{\xi_n(n): n \in \mathbb{N}\}$.]

12. (a) Suppose $\chi(x,y)$ is Γ. Show that there is a Γ formula $\xi(x)$ such that
(i) $T \vdash \forall x(\xi(x) \to \chi(x,m))$ for all m,
(ii) if $\eta(x)$ is Γ^d and $T \vdash \forall x(\xi(x) \to \eta(x))$, there is a q such that
$T \vdash \forall x(\forall y \leqslant q \chi(x,y) \to \eta(x))$

(compare Lemma 2).

(b) Let $\{\gamma_k(x): k \in N\}$ be an r.e. set of Γ formulas. Show that there is a Γ formula $\eta(x)$ such that for every k, $T \vdash \forall x(\eta(x) \to \gamma_k(x))$ and for every Γ^d formula $\chi(x)$, if $T \vdash \forall x(\eta(x) \to \chi(x))$, there is an m such that $T \vdash \forall x(\wedge\{\gamma_k(x): k \leq m\} \to \chi(x))$ (compare Theorem 4).

(c) Show that there is a Γ formula $\xi(x)$ such that $T \vdash \xi(n)$ for every n and if $\eta(x)$ is Γ^d and $T \vdash \forall x(\eta(x) \to \xi(x))$, there is an n such that $T \vdash \forall x(\eta(x) \to x \leq n)$.

(d) Let $\{\gamma_k(x): k \in N\}$ be an r.e. set of Γ formulas and let $Y = \{k: \exists m T \vdash \gamma_0(k) \vee \ldots \vee \gamma_m(k)\}$. Show that there is a Γ formula $\eta(x)$ such that for every k, $T \vdash \forall x(\gamma_k(x) \to \eta(x))$ and $\eta(x)$ numerates Y in T.

13. Corollary 2 can be generalized as follows. Let X be a set of ordered pairs (F,G) of formulas and suppose there is an interpretation I such that $F^I \dashv_{\Pi_n} G^I$ iff $(F,G) \in X$. Then X satisfies the following conditions:
(i) if $F \to G$ is a tautology, then $(G,F) \in X$,
(ii) if $(F,G) \in X$ and $(G,H) \in X$, then $(F,H) \in X$,
(iii) if $(F,G) \in X$ and $(F,H) \in X$, then $(F, G \vee H) \in X$,
(iv) $(\bot, \top) \notin X$.

Show that for any X satisfying these conditions, there is a model **M** for Π-conservativity such that for all F, G. $\Vdash (F,G)$ iff $(F,G) \in X$. (Thus, if there is an interpretation I such that $F^I \dashv_{\Pi_n} G^I$ iff $(F,G) \in X$, this can be derived from Theorem 7.) [Hint: For every truth-value assignment t, let E_t be the conjunction of variables and negations of variables "describing" t. Let F_j, $j = 0, 1, \ldots$, be those formulas F for which $(F, \top) \notin X$. Let V_j be defined by: $t \in V_j$ iff $(F_j, E_t) \notin X$. Then for all F, G, $(F,G) \in X$ iff for every j and every $t \in V_j$, if $t(G) = $ true, there is a $t' \in V_j$ such that $t'(F) = $ true.]

14. Derive Theorem 5(a) from Theorem 8.

In Exercises 15, 16, 17 let
$$\chi_n(x,y,z) := \text{rank}_n(x,z) \leq z \wedge \forall uv \leq \text{rank}_n(x,z) \exists w \leq z\{\Sigma_n\}(\lambda(\dot{y}),u,v,w).$$

15. Let $f \in \{0,1,2\}^N$ be such that, provably in PA:
Stage 0. $f(0) = 0$.
Stage s+1. Suppose $f(s) = a$.
   Case 1. $a \leq 1$ and $\chi_n(a+1,a,s)$.
      $f(s+1) = a+1$.
   Case 2. Otherwise.

$f(s+1) = a$.

Let $\varphi_0 := \lambda(0)$ and $\varphi_1 := \lambda(0) \vee \lambda(1)$. Then φ_0 and φ_1 are Π_n and $\vdash \varphi_0 \to \varphi_1$. Show that $T \nvdash \neg \varphi_0$, $\neg \varphi_0$ is Π_n-conservative over $T + \varphi_1$, φ_1 is Σ_n-conservative over $T + \neg \varphi_0$, and $\neg \varphi_1$ is Π_n-conservative over T.

16. Let $\{c_0, c_1, d\} = \{1, 2, 3\}$. Let f be such that, provably in PA:

Stage 0. $f(0) = 0$.

Stage s+1. Suppose $f(s) = a$.

Case 1. (i) Either $a = 0$ and $b = c_i$ or $a = c_i$ and $b = d$ and (ii) $\chi_n(b,a,s)$. Let b be the least number for which this holds.

$f(s+1) = b$.

Case 2. Otherwise.

$f(s+1) = a$.

Show that $\lambda(c_0) \vee \lambda(c_1) \dashv_{\Sigma_n} \lambda(d)$ and $\lambda(c_i) \not\dashv_{\Sigma_n} \lambda(d)$, $i = 0, 1$.

17. Let $\{c_0, c_1, d_0, d_1\} = \{1, 2, 3, 4\}$. Define f in such a way that $\lambda(d_i) \dashv_{\Pi_n} \lambda(c_i)$ $\dashv_{\Pi_n} \lambda(0)$, $\lambda(c_i) \dashv_{\Sigma_n} \lambda(d_i)$, and $\lambda(0) \vee \lambda(c_i) \vee \lambda(d_i)$ and $\lambda(0) \vee \lambda(c_0) \vee \lambda(c_1)$ are Π_n, $i = 0, 1$. [Hint: Let f be such that:

Stage 0. $f(0) = 0$.

Stage s+1. Suppose $f(s) = a$.

Case 1. (i) Either $a = 0$ and $b = c_i$ or $a = c_i$ and $b = d_i$ and (ii) $\chi_n(b,a,s)$. Let b be the least number for which this holds.

$f(s+1) = b$.

Case 2. Otherwise.

$f(s+1) = a$.]

Notes for Chapter 5

The general concept Γ-*conservative* is due to Guaspari (1979). Theorem 1 is due to Kreisel (1962). Lemma 2 is due to Lindström (1984a). Lemma 3 and Theorem 2 with $X = Th(T)$ are due to Guaspari (1979); for somewhat stronger results, also due to Guaspari (1979), see Exercise 5(b). The proofs of Lemma 3 and Theorem 2 are from Lindström (1984a). Lemma 4 is due to Lindström (1984a). (Lemmas 2 and 4 and their proofs are similar to and were inspired by results of Guaspari (1979), Solovay (cf. Guaspari (1979)), and Hájek (1971); for further applications, see e.g. Hájek and Pudlák (1993).) Theorem 3 less the references to the set X is due to Solovay (cf. Guaspari (1979)); see also Jensen and Ehrenfeucht (1976); the full result is proved in Smoryński (1981a) and in Lindström (1984a), in the former as an application of a very general construction another special case of which is the fixed point mentioned in Exercise 3.7(a). Theorem 4 is due to Lindström (1984a). Lemma 6, Theorem 5, and Corollary 1 are due to Bennet (1986), (1986a); Theorem 5(a) can also be derived from Theorem 8. Corollary 1 with Σ_n replaced by Π_n is false (Exercise 3). (It follows that the lattices of Σ_m and Π_n sentences, modulo provable equivalence in T, are not isomorphic.) Theorem 6 is due jointly to Solovay (cf. Hájek (1979)) and Hájek (1979), Quinsey (1980), (1981), and Lindström (1984a); the present proof is due to Lindström (1984a); Quinsey also shows that $\Pi_1 \cap \mathrm{Cons}(\Sigma_1,T)$ is complete Π_2^0 if T is not Σ_1-sound; cf. also Bennet (1986), (1986a). Lemma 8 is due to Lindström (1993).

Theorem 7 is essentially a special case of results of Hájek and Montagna (1990) (cf. also Japaridze (1994)), for n = 1, and Ignatiev (1991) for n > 1; Ignatiev (1991) also has similar results for Σ_n with n ⩾ 3, but his proofs are more complicated and will not be presented here; there are no corresponding results for Σ_1 and Σ_2. (These results belong to the logic of provability, a subject not treated in this book; cf. e.g. Boolos (1979), (1993), Japaridze and de Jongh (1998), Lindström (1996), Smoryński (1985).) The first Solovay function and the associated interpretation were defined in Solovay (1976). Our models for Π-conservativity were obtained by simplifying the so called Veltman models, defined by Veltman (cf. de Jongh and Veltman (1990) and Japaridze and de Jongh (1998)). The first rank functions were introduced by Berarducci (1990) and Shavrukov (1988). For a generalization of Corollary 2, see Exercise 13. For further examples of Solovay functions, including Solovay's original definition, see Exercises 15, 16, 17, Boolos (1979), (1993), Japaridze and de Jongh (1998), Lindström (1996), Smoryński (1985), and the literature referred to in these works.

Exercise 2(a) is due to Smoryński (1980), (1981b); Exercise 2(b) is due to Švejdar (cf. Hájek and Pudlák (1993)). Exercise 4 is due to Kent (1973). Exercise 5(b) is due to Guaspari (1979). Exercise 7(a) is due to Smoryński (1981a). Exercise 9 is due to Misercque (1983). What is essentially Exercise 13 is proved in Strannegård (1997) as a special case of a much more general result. From Theorem 7 and Exercise 13 it follows that whether or not for a given X there is an interpretation I such that $F^I \dashv_{\Pi_n} G^I$ iff (F,G) ∈ X is independent of n and T. Exercise 17, for n = 1, is (a lemma) due to Shavrukov (with a different proof, unpublished); what is essentially the full result was proved by Bennet (in a different way, unpublished).

6. INTERPRETABILITY

Let S and S' be arbitrary theories. S' is interpretable in S if, roughly speaking, the primitive concepts and the range of the variables of S' are definable in S in such a way as to turn every theorem of S' into a theorem of S. If, in addition every nontheorem of S' is transformed into a nontheorem of S, then S' is faithfully interpretable in S.

In this chapter, we assume that PA ⊣ T. Thus, T is essentially reflexive.

§1. Interpretability. Let S and S' be arbitrary theories. By a *translation* (of the language of S' into the language of S) we understand a function t on the set of formulas (of S') into the set of formulas (of S) for which there are formulas $\eta_0(x)$, $\eta_S(x,y)$, $\eta_+(x,y,z)$, $\eta_\times(x,y,z)$ and a formula $\mu_t(x)$ such that t satisfies the following conditions for all formulas $\varphi, \psi, \xi(x)$:

(*) $t(x = y) := x = y,$
  $t(x = 0) := \eta_0(x),$
  $t(Sx = y) := \eta_S(x,y),$
  $t(x + y = z) := \eta_+(x,y,z),$
  $t(x \times y = z) := \eta_\times(x,y,z),$
  $t(\neg\varphi) := \neg t(\varphi),$
  $t(\varphi \wedge \psi) := t(\varphi) \wedge t(\psi),$
  $t(\exists x \xi(x)) := \exists x(\mu_t(x) \wedge t(\xi(x))).$

(Here x, y, z are arbitrary variables.) We assume that $\forall$ and the connectives $\vee, \rightarrow, \leftrightarrow$ are defined in terms of $\exists, \neg, \wedge$. Note that t, on the formulas for which it is defined by the above conditions, is uniquely determined by its values on atomic formulas together with the formula $\mu_t(x)$.

So far $t(\varphi)$ is only defined provided that every atomic formula in φ contains at most one arithmetical symbol. But every formula φ is logically equivalent to a formula φ^* having this property and obtained from φ in some canonical way. We may then let $t(\varphi) := t(\varphi^*)$. Clearly t is a primitive recursive function.

The translation t is an *interpretation in* S iff
(**) $S \vdash \exists x \mu_t(x),$
  $S \vdash \exists x(\mu_t(x) \wedge \forall y(\mu_t(y) \rightarrow (\eta_0(y) \leftrightarrow y = x))),$
  $S \vdash \forall x(\mu_t(x) \rightarrow \exists y(\mu_t(y) \wedge \forall z(\mu_t(z) \rightarrow (\eta_S(x,z) \leftrightarrow z = y)))),$
  $S \vdash \forall xy(\mu_t(x) \wedge \mu_t(y) \rightarrow$
    $\exists z(\mu_t(z) \wedge \forall u(\mu_t(u) \rightarrow (\eta_*(x,y,u) \leftrightarrow u = z)))), * = +, \times.$

§1. Interpretability

Thus, t is an interpretation in S iff $S \vdash t(\varphi)$ for every logically valid sentence φ.

t is an *interpretation of* S' *in* S, t: $S' \leqslant S$, iff $S \vdash t(\varphi)$ for every φ such that $S' \vdash \varphi$. S' is *interpretable in* S, $S' \leqslant S$, if there is an interpretation of S' in S. $S' < S$ means that $S' \leqslant S \not\leqslant S'$.

Trivially, if $S' \dashv S$, then $S' \leqslant S$. The reader should check that $\leqslant$ is a transitive relation. Also note that if $S' \leqslant S$, then every finite subtheory of S' is interpretable in a finite subtheory of S.

If $S' \leqslant S$ and S is consistent, so is S'. For suppose S' is not consistent. Let φ be any sentence. Then $S' \vdash \varphi \wedge \neg\varphi$. But then $S \vdash t(\varphi \wedge \neg\varphi)$. But $t(\varphi \wedge \neg\varphi) := t(\varphi) \wedge \neg t(\varphi)$, whence $S \vdash t(\varphi) \wedge \neg t(\varphi)$ and so S is inconsistent.

Since every translation t is a primitive recursive function, we may in (extensions of) PA use t as a function symbol. t can always be defined such that the following Fact holds and the argument in the preceding paragraph can be formalized in PA.

Fact 12. Suppose t: $S' \leqslant S$.
   (a) The conditions (*) and (**) are provable in PA.
   (b) $PA \vdash Pr_\emptyset(x) \to Pr_S(t(x))$.

This Fact has the following:

Corollary 1. Suppose t: $S' \leqslant S$ and S' is finite. Then $PA \vdash Pr_{S'}(x) \to Pr_S(t(x))$ and consequently $PA \vdash Con_S \to Con_{S'}$.

The assumption that S' is finite in Corollary 1 cannot be omitted: $S' \leqslant S$ may be true but not provable in PA (see Corollary 5 and Theorem 12, below). But we do have the following weaker result. (Recall that a *numeration* of a set X numerates X in PA.)

Theorem 1. Suppose $S_0 \leqslant S_1$ and let $\sigma_1(x)$ be a Σ_1 numeration of S_1. There is then a Σ_1 numeration $\sigma_0(x)$ of S_0 such that
$$PA \vdash Con_{\sigma_1} \to Con_{\sigma_0}.$$

Proof. Suppose t: $S_0 \leqslant S_1$. Let $\sigma(x)$ be a PR binumeration of S_0 and let $\sigma_0(x) := \sigma(x) \wedge Pr_{\sigma_1}(t(x))$. Then $\sigma_0(x)$ is a Σ_1 numeration of S_0 and
(1) $PA \vdash Pr_{\sigma_0}(x) \to Pr_{\sigma_1}(t(x))$.
To prove this, we reason (informally) in PA as follows: "Suppose φ is derivable from formulas satsifying $\sigma_0(x)$. Then there are $\psi_0,\ldots,\psi_n$ of formulas

satisfying $\sigma_0(x)$, such that $\bigwedge\{\psi_k: k \leq n\} \to \varphi$ is provable in logic. But then, by Fact 12 (this chapter), $t(\bigwedge\{\psi_k: k \leq n\}) \to t(\varphi)$ is provable from the set defined by $\sigma_1(x)$. But $t(\bigwedge\{\psi_k: k \leq n\}) := \bigwedge\{t(\psi_k): k \leq n\}$. Also, by the definition of $\sigma_0(x)$, each $t(\psi_k)$ is derivable from the set defined by $\sigma_1(x)$. But then so is $\bigwedge\{t(\psi_k): k \leq n\}$. It follows that $t(\varphi)$ is derivable from the set defined by $\sigma_1(x)$." This proves (1).

From (1) we easily get the desired conclusion. ∎

Theorem 1 in combination with Gödel's second incompleteness theorem (Theorem 2.4) yields the following strengthening of Gödel's result. For a different improvement of Theorem 2.4, see Theorem 8, below.

Theorem 2. (a) $T + \text{Con}_T \not\leq T$.

(b) If $\tau(x)$ is any Σ_1 formula numerating (an extension of) T in T, then $T + \text{Con}_\tau \not\leq T$.

Proof. (a) Suppose $T + \text{Con}_T \leq T$. Then, by Theorem 1, there is a Σ_1 numeration $\tau'(x)$ of $T + \text{Con}_T$ such that $T + \text{Con}_T \vdash \text{Con}_{\tau'}$. By Theorem 2.4 it now follows that $T + \text{Con}_T$ is inconsistent. But then, since $T + \text{Con}_T \leq T$, T is inconsistent, contrary to Convention 2.

(b) Similar. ∎

Since Con_T is Π_1, Theorem 2 is also a direct consequence of Theorem 2.4 and the following:

Lemma 1. If π is a Π_1 sentence and $Q + \pi \leq T$, then $T \vdash \pi$.

Proof. There is a k such that $Q + \pi \leq T|k$. So, by Corollary 1, $T \vdash \text{Con}_{T|k} \to \text{Con}_{Q+\pi}$. It follows that $T \vdash \text{Con}_{Q+\pi}$. Since $\neg\pi$ is Σ_1, we have, by provable Σ_1-completeness, $T \vdash \neg\pi \to \neg\text{Con}_{Q+\pi}$. It follows that $T \vdash \pi$. ∎

Note that we have actually proved that $Q + \text{Con}_T \not\leq T$.

In Chapter 2 (Corollary 2.1) we proved that PA is essentially infinite (in fact, PA is essentially unbounded; Corollary 4.1). This can now be improved as follows:

Theorem 3. T is not interpretable in any finite subtheory of T.

Proof. Let S be a finite subtheory of T and suppose $T \leq S$. By Theorem 1, there is then a Σ_1 numeration $\tau(x)$ of T such that $\text{PA} \vdash \text{Con}_S \to \text{Con}_\tau$. Since, by Fact 11, T is reflexive, we have $T \vdash \text{Con}_S$ and so $T \vdash \text{Con}_\tau$, contradicting Theorem 2.4. ∎

§1. Interpretability

Most positive results on the existence of interpretations in the sequel are applications of the following fundamental result, the arithmetization of Gödel's completeness theorem.

Theorem 4. Let $\sigma(x)$ be a formula numerating S in T. Then $S \leqslant T + \text{Con}_\sigma$.

Proof (informal outline). A full proof of this result would be quite long and we shall be content to give a fairly detailed sketch. The main idea is to show that (the denumerable case of) the Henkin completeness proof for first order logic can be formalized in PA. (The reader is assumed to be familiar with that proof.)

We begin with an outline of Henkin's proof. Let S be a (countable) set of sentences (theory) assumed to be consistent. Let c_n, $n \in \mathbb{N}$, be new individual constants. Let L be the language obtained from L_S by adding the constants c_n. Let $\alpha_n(x_n)$, $n \in \mathbb{N}$, be a primitive recursive enumeration of all formulas of L with one free variable. We can then form a primitive recursive set
$$Z = \{\exists x_n \alpha_n(x_n) \to \alpha_n(c_{jn}) : n \in \mathbb{N}\}$$
such that
(1)   for every sentence θ of S, if $S + Z \vdash \theta$, then $S \vdash \theta$.
It follows that $S + Z$ is consistent.

Now let θ_n, $n \in \mathbb{N}$, be a primitive recursive enumeration of all sentences of L. The sentences φ_n are then inductively defined as follows:
(2) $\varphi_n = \theta_n$ if $S + Z \vdash \bigwedge\{\varphi_m : m < n\} \to \theta_n$,
      $= \neg \theta_n$ otherwise.
(Here $\bigwedge\{\varphi_m : m < 0\} := \top$.) φ_n is not in general a recursive function of n.

Let $X = \{\varphi_n : n \in \mathbb{N}\}$. Then
(3) $\text{Th}(S) \subseteq X$
and, since $S + Z$ is consistent,
(4)   X is *Henkin complete*
in the sense that X is complete and consistent and for every formula $\alpha(x)$ of L with the one free variable x, if $\exists x \alpha(x) \in X$, there is a constant c_k such that $\alpha(c_k) \in X$.

We can now define a model
$$\mathbf{M} = (M, S^M, +^M, \times^M, 0^M)$$
of X in the following way. The domain M of the model is the set $\{c_n : n \in \mathbb{N}\}$. (Here we ignore the minor difficulty that X may contain sentences of the form $c_k = c_m$ with $k \neq m$ and so the members of M cannot in general be the constants themselves but must instead be certain "equivalence classes" of these constants or, in the present context, members of such equivalence

classes. If we disregard the trivial case where S has only finite models, this can be avoided by defining Z in a slightly different way.)

$0^M = c_{i_0}$, $c_n^M = c_n$,
$S^M = \{(c_k, c_m) : Sc_k = c_m \in X\}$,
$+^M = \{(c_k, c_m, c_n) : c_k + c_m = c_n \in X\}$,
$\times^M = \{(c_k, c_m, c_n) : c_k \times c_m = c_n \in X\}$,

where c_{i_0} is the (uniquely determined) constant such that $0 = c_{i_0} \in X$.

Finally, it can be shown, by induction and using the fact that X is Henkin complete, that for every sentence φ of L,

(5) φ is true in **M** iff $\varphi \in X$.

This is true, by the definition of **M**, if φ is atomic.

Finally, Th(S) $\subseteq$ X and so **M** is a model of Th(S).

We can now transform this into a proof that $S \leqslant T + \text{Con}_\sigma$ in the following way. We first define in PA a primitive recursive function $c(x)$ (= the xth new individual constant). By a c-formula we understand a formula obtained from a formula of L_A by replacing each free variable v by $c(\dot{v})$. (Thus, the c-formulas are the counterparts of the sentences of L.) Let $\zeta(x)$ be a suitably defined PR binumeration of Z, where Z is defined as above except that we now use the function symbol c. Then (the reader will hopefully believe that) for every sentence φ of S,

(6) $PA \vdash \text{Pr}_{\sigma \vee \zeta}(\varphi) \to \text{Pr}_\sigma(\varphi)$

(compare (1)). It follows that

(7) $PA \vdash \text{Con}_\sigma \to \text{Con}_{\sigma \vee \zeta}$.

The inductive definition of φ_n can, using methods available in PA, be turned into an explicit definition. Let $\chi(x,y)$ be a suitable formalization of this explicit definition (cf. Chapter 1, p. 11). Let $\xi(x) := \exists y \chi(x,y)$. (Thus, intuitively, $\xi(x)$ means "x is a member of X".) Then (compare (3))

(8) $PA \vdash \text{Pr}_\sigma(x) \to \xi(x)$.

Let Hcm_ξ be the sentence saying that the set defined by $\xi(x)$ is Henkin complete. Thus, for all c-formulas α, β,

(9) $PA + \text{Hcm}_\xi \vdash \xi(\neg \alpha) \leftrightarrow \neg \xi(\alpha)$.
(10) $PA + \text{Hcm}_\xi \vdash \xi(\alpha) \wedge \text{Pr}_\emptyset(\alpha \to \beta) \to \xi(\beta)$.

Moreover, for every formula $\alpha(x)$ such that $\exists x \alpha(x)$ is a c-formula,

(11) $PA + \text{Hcm}_\xi \vdash \xi(\exists x \alpha(x)) \to \exists u \xi(\alpha(c(\dot{u})))$.

The (inductive) proof of (4) does not use any means of proof beyond those available in PA. Thus, we get $PA \vdash \text{Con}_{\sigma \vee \zeta} \to \text{Hcm}_\xi$ and so, by (7),

(12) $PA \vdash \text{Con}_\sigma \to \text{Hcm}_\xi$.

We can now define a translation t, corresponding to the model **M**, as follows. Let

§1. *Interpretability*

$$\mu_t(x) := \exists u(x = c(u)),$$
$$t(x = 0) := \exists u(x = c(u) \wedge \xi(0 = c(\dot{u}))),$$
$$t(Sx = y) := \exists uv(x = c(u) \wedge y = c(v) \wedge \xi(Sc(\dot{u}) = c(\dot{v}))),$$
$$t(x * y = z) := \exists uvw(x = c(u) \wedge y = c(v) \wedge z = c(w) \wedge$$
$$\xi(c(\dot{u}) * c(\dot{v}) = c(\dot{w}))), * = +, \times$$

These equations uniquely determine t.

The proof corresponding to the proof of (5) now yields for every formula $\beta(x_0, \ldots, x_{n-1})$ of L_A containing no free variables other than $x_0, \ldots, x_{n-1}$,

(13) $PA + Hcm_\xi \vdash \mu_t(x_0) \wedge \ldots \wedge \mu_t(x_{n-1}) \to (t(\beta(x_0, \ldots, x_{n-1})) \leftrightarrow$
$\exists u_0, \ldots, u_{n-1}(x_0 = c(u_0) \wedge \ldots \wedge x_{n-1} = c(u_{n-1}) \wedge \xi(\beta(c(\dot{u}_0), \ldots, c(\dot{u}_{n-1})))).$

By the definition of t, this holds for atomic $\beta(x_0, \ldots, x_{n-1})$. The inductive steps dealing with $\neg$ and $\wedge$ follow easily, by (9) and (10).

Let us consider the step dealing with $\exists$. For simplicity, let $n = 1$ and write x for x_0. Let $\alpha(x,y)$ be such that $\beta(x) := \exists y\alpha(x,y)$. Then $t(\beta(x)) := \exists y(\mu_t(y) \wedge t(\alpha(x,y)))$. By the inductive hypothesis,

$$PA + Hcm_\xi \vdash \mu_t(x) \wedge \mu_t(y) \to (t(\alpha(x,y)) \leftrightarrow$$
$$\exists uv(x = c(u) \wedge y = c(v) \wedge \xi(\alpha(c(\dot{u}),c(\dot{v}))))).$$

By (10) and (11),

$$PA + Hcm_\xi \vdash \exists v\xi(\alpha(c(\dot{u}),c(\dot{v}))) \leftrightarrow \xi(\exists y\alpha(c(\dot{u}),y)).$$

But then it is fairly easy to see that

$$PA + Hcm_\xi \vdash \mu_t(x) \to (\exists y(\mu_t(y) \wedge t(\alpha(x,y))) \leftrightarrow$$
$$\exists u(x = c(u) \wedge \xi(\exists y\alpha(c(\dot{u}),y)))),$$

as desired. This proves (13).

From (12) and (13), we get for every sentence φ,

(14) $PA + Con_\sigma \vdash t(\varphi) \leftrightarrow \xi(\varphi).$

Finally, let φ be any sentence provable in S. Then $T \vdash Pr_\sigma(\varphi)$. Hence, by (8), $T \vdash \xi(\varphi)$ and so, by (14), $T + Con_\sigma \vdash t(\varphi)$. It follows that $t: S \leqslant T + Con_\sigma$.

This concludes our sketch of the proof of Theorem 4. ∎

If we don't insist on mimicking every detail of Henkin's proof, we can instead use the simpler interpretation t' defined in the following way:

$$\mu_{t'}(x) := x = x,$$
$$t'(x = 0) := \xi(0 = c(\dot{x})),$$
$$t'(Sx = y) := \xi(Sc(\dot{x}) = c(\dot{y})),$$
$$t'(x + y = z) := \xi(c(\dot{x}) + c(\dot{y}) = c(\dot{z})),$$
$$t'(x \times y = z) := \xi(c(\dot{x}) \times c(\dot{y}) = c(\dot{z})).$$

Thus, $\mu_{t'}(x)$ is trivial and can be omitted. (This is true as long as we are dealing with theories of (elementary) arithmetic; it is *not* true in general.)

It is via the following lemma, the (Feferman–)Orey–Hájek Lemma (and Theorem 6, below) that Theorem 4 becomes such a powerful tool in the theory of interpretability (of arithmetical theories; see also Lemma 8.4).

Lemma 2. $S \leqslant T$ iff $T \vdash \mathrm{Con}_{S|k}$ for every k.

To prove this we need the following lemma whose proof is essentially the same as that of Theorem 2.7.

Lemma 3. Suppose $T \vdash \mathrm{Con}_{S|k}$ for every k. Let $\sigma(x)$ be any formula binumerating S in T and let
$$\sigma^*(x) := \sigma(x) \wedge \mathrm{Con}_{\sigma|x}.$$
Then (i) $\sigma^*(x)$ binumerates S in T and (ii) $PA \vdash \mathrm{Con}_{\sigma^*}$.

Proof of Lemma 2. Suppose first $S \leqslant T$. Let k be arbitrary. There is then an m such that $S|k \leqslant T|m$. By Corollary 1, $PA \vdash \mathrm{Con}_{T|m} \to \mathrm{Con}_{S|k}$. But $T \vdash \mathrm{Con}_{T|m}$ and so $T \vdash \mathrm{Con}_{S|k}$.

Next suppose $T \vdash \mathrm{Con}_{S|k}$ for every k. Let $\sigma(x)$ be a PR binumeration of S and let $\sigma^*(x) := \sigma(x) \wedge \mathrm{Con}_{\sigma|x}$. Then, by Lemma 3, $\sigma^*(x)$ binumerates S in T and $PA \vdash \mathrm{Con}_{\sigma^*}$. Hence, by Theorem 4, $S \leqslant T$. ∎

There are alternative notions of interpretability more general than the one defined here. For example, we may "interpret" the equality symbol = of one theory S as a certain relation definable in another S' (and having, provably in S', the required properties) or we may "interpret" the individuals of S as finite sequences of individuals of S' etc. It turns out, however, that if S is "interpretable" in T in some such more general, and reasonably natural, sense, then, by Lemma 2, $S \leqslant T$ (and conversely). Thus, in the present context, there is no reason to consider these more general "interpretations".

From Lemmas 2 and 3 and Theorem 4 we get the following:

Corollary 2. $S \leqslant T$ iff there is a formula $\sigma(x)$ (bi)numerating S in T such that $T \vdash \mathrm{Con}_\sigma$.

From Lemma 2 we also obtain the following result known as Orey's compactness theorem.

Theorem 5. $S \leqslant T$ iff $S|k \leqslant T$ for every k.

In the following we use A, B, etc. to denote (consistent, primitive recursive) extensions of T. Recall that $A \dashv_\Gamma B$ means that every Γ sentence provable in A is provable in B.

Theorem 6. $A \leqslant B$ iff $A \dashv_{\Pi_1} B$.

Proof. Suppose first $A \dashv_{\Pi_1} B$. Now, $A \vdash \text{Con}_{A|k}$ for every k. It follows that $B \vdash \text{Con}_{A|k}$ for every k. But then, by Lemma 2, $A \leqslant B$.
Suppose next $A \leqslant B$. Let π be any Π_1 sentence such that $A \vdash \pi$. By Lemma 1, $B \vdash \pi$, as desired. ∎

By Theorem 6, $A + \varphi \leqslant A$ iff φ is Π_1-conservative over A.
Theorem 6 has the following immediate:

Corollary 3. If $A \leqslant B$ and σ is any Σ_1 sentence, then $A + \sigma \leqslant B + \sigma$.

Combining Theorem 6 and Theorem 4.5 we get:

Corollary 4. $T + \text{Rfn}_T \leqslant PA + \text{Con}_T^\omega$.

In fact, this follows directly from Lemma 2 and the fact, established in the proof of Theorem 4.5, that $PA + \text{Con}_T^\omega \vdash \text{Con}_{T_n}$ for every n.
Theorem 6 can also be used to prove the following model-theoretic characterization of interpretability:

Theorem 7. $A \leqslant B$ iff for every model **M** of B, there is a model **M'** of A such that **M** is (isomorphic to) an initial segment of **M'**.

Proof (sketch). "If". Let θ be any Π_1 sentence such that $A \vdash \theta$. We show that θ holds in all models of B. Let **M** be any model of B. By hypothesis, there is a model **M'** of A such that **M** is isomorphic to an initial segment of **M'**. θ holds in **M'**. Since θ is Π_1, it follows that θ holds in **M**. Thus, θ holds in all models of B and so $B \vdash \theta$. We have shown that $A \dashv_{\Pi_1} B$ and so $A \leqslant B$, by Theorem 6.

"Only if". Let $t: A \leqslant B$. Let **M** be any model of B and let **M'** be the structure defined by t in **M**. **M'** is a model of A. Since induction holds in **M**, we can in **M** define a function f on M satisfying the following conditions: $f(0^M) = 0^{M'}$, $f(S^M(a)) = S^{M'}(f(a))$. f maps **M** isomorphically onto an initial segment of **M'**. ∎

Given Theorem 6, we can now derive Theorems 8–12 below (and many similar results, which will not be explicitly stated) as corollaries to results (and Exercises) from Chapter 5.

Like Theorem 2 the following result is a sharpening of Gödel's second incompleteness theorem.

Theorem 8. $T + \neg \text{Con}_T \leqslant T$.

Proof. This follows from Theorem 5.1 and Theorem 6. ∎

A more direct proof of Theorem 8 is as follows. We need the following:

Lemma 4. If $S \leqslant S' + \varphi_0$ and $S \leqslant S' + \varphi_1$, then $S \leqslant S' + \varphi_0 \vee \varphi_1$. Thus, if $S + \varphi \leqslant S + \neg\varphi$, then $S + \varphi \leqslant S$.

Proof. Suppose $t_i: S \leqslant S' + \varphi_i$, $i = 0, 1$. Let t be the translation which coincides with t_0 if φ_0 and with t_1 if $\neg\varphi_0 \wedge \varphi_1$. Thus, for example,
$$\mu_t(x) := (\varphi_0 \wedge \mu_{t_0}(x)) \vee (\neg\varphi_0 \wedge \varphi_1 \wedge \mu_{t_1}(x)).$$
It follows that for all φ,
(1) $S' + \varphi_0 \vdash t(\varphi) \leftrightarrow t_0(\varphi)$,
(2) $S' + \neg\varphi_0 \wedge \varphi_1 \vdash t(\varphi) \leftrightarrow t_1(\varphi)$.
Now, suppose $S \vdash \varphi$. Then $S' + \varphi_0 \vdash t_0(\varphi)$ and so, by (1), $S' + \varphi_0 \vdash t(\varphi)$. Also $S' + \varphi_1 \vdash t_1(\varphi)$ and so, by (2), $S' + \neg\varphi_0 \wedge \varphi_1 \vdash t(\varphi)$. It follows that $S' + \varphi_0 \vee \varphi_1 \vdash t(\varphi)$. Thus, $t: S \leqslant S' + \varphi_0 \vee \varphi_1$, as desired. ∎

By Corollary 2.2, $T + \text{Con}_T \vdash \text{Con}_{T + \neg \text{Con}_T}$. But then, by Theorem 4, $T + \neg \text{Con}_T \leqslant T + \text{Con}_T$ and so, by Lemma 4, $T + \neg \text{Con}_T \leqslant T$, as desired. (In this proof of Theorem 8 it is not necessary to assume that T is (essentially) reflexive.)

Theorem 9. Suppose X is r.e. and monoconsistent with T. There is then a Σ_1 sentence φ such that $T + \varphi \leqslant T$ and $\varphi \notin X$.

Proof. This follows from Theorem 5.2 and Theorem 6. ∎

Theorem 9 has the following immediate:

Corollary 5. (a) The set $\{\varphi \in \Sigma_1 : T + \varphi \leqslant T\}$ is not r.e.
(b) Let $\tau(x)$ be a formula numerating T in T such that $T \nvdash \neg \text{Con}_\tau$. There is then a ($\Sigma_1$) sentence φ such that $T + \varphi \leqslant T$ and $T \nvdash \text{Con}_\tau \rightarrow \text{Con}_{\tau + \varphi}$.

Theorem 10. Suppose X is r.e. and monoconsistent with T. There is then a sentence φ such that $T + \varphi \leqslant T$, $T + \neg\varphi \leqslant T$, $\varphi \notin X$, $\neg\varphi \notin X$.

Proof. By Theorem 5.3 we can take φ to be, say, a Σ_2 sentence such that φ is

§1. Interpretability

Π_2-conservative and $\neg\varphi$ is Σ_2-conservative over T. Now use Theorem 6. ∎

A sentence φ such that $T + \varphi \leqslant T$, $T + \neg\varphi \leqslant T$ is known as an *Orey sentence* for T. Clearly, any Orey sentence for T is undecidable in T.

Theorems 9 and 10 can also be applied as follows. There are consistent finitely axiomatized extensions U of T in languages extending L_A. In fact, U may chosen to be a conservative extension of T in the sense that for every sentence φ of L_A, $U \vdash \varphi$ iff $T \vdash \varphi$. Thus, U and T are equivalent in terms of provability of sentences of L_A. So it is natural to ask if U and T are (ever) equivalent in terms of interpretability of sentences of L_A in the sense that for every sentence φ of L_A, $T + \varphi \leqslant T$ iff $U + \varphi \leqslant U$. (We assume the reader can extend the defintion of "interpretation" and "interpretable in" to the case where the theories need not be formalized in L_A.) The answer is a resounding "no" (see also Corollary 8.8). To prove this we need the following essentially trivial lemma whose proof is left to the reader.

Lemma 5. Let V be any r.e. theory, not necessarily in L_A. Then the set $\{\varphi: U + \varphi \leqslant V\}$ is r.e.

From Theorem 9 and Lemma 5 we get:

Corollary 6. Let U, V be as above. There is then a Σ_1 sentence φ such that $T + \varphi \leqslant T$ and $U + \varphi \not\leqslant V$.

Similarly, Theorem 10 and Lemma 5 jointly imply:

Corollary 7. Let U, V be as above. There is then a sentence φ such that $T + \varphi \leqslant T$, $T + \neg\varphi \leqslant T$, $U + \varphi \not\leqslant V$, and $U + \neg\varphi \not\leqslant V$.

As we saw in Chapter 4, speaking in terms of provability, we have to distinguish between finite, infinite, and unbounded extensions of a given theory T. In terms of interpretability the situation is quite different. We write $S \equiv S'$ to mean that $S \leqslant S' \leqslant S$.

Theorem 11. (a) If $A \dashv B$, there is a sentence φ such that $A + \varphi \equiv B$.

(b) Let X be an r.e. set of Σ_1 sentences. Then there is a Σ_1 sentence σ such that $T + \sigma \equiv T + X$.

Proof. (a) Let $X = \text{Th}(B) \cap \Pi_1$. Then, by Theorem 6, $A + X \equiv B$. By Theorem 5.4(a), there is a sentence φ such that $A + \varphi$ is a Π_1- conservative extension of $A + X$. By Theorem 6, $A + \varphi \equiv A + X$ and so $A + \varphi \equiv B$. ♦

(b) This follows from Theorems 5.4(a) and 6. ∎

Finally, we have a result which proves the claim made earlier that the fact that, for example, $A + \varphi \leq B$ does not imply that this is provable in PA, or in any other preassigned consistent axiomatizable theory.

From the definition of $\leq$ it is clear that the set $\{\varphi: A + \varphi \leq B\}$ is Σ_3^0. From Theorem 6, it follows, however, that $\{\varphi: A + \varphi \leq B\}$ is Π_2^0. That this cannot be improved follows from:

Theorem 12. Suppose $A \leq B$. Then the set $\{\varphi \in \Sigma_1: A + \varphi \leq B\}$ is a complete Π_2^0 set.

Proof. For $A = B$, this follows from Theorem 5.6 and Theorem 6; we leave the proof of the general case to the reader. ∎

A translation t is given by a finite amount of information which can certainly be coded by a natural number; thus we may "identify" t with that number. Let $\text{Int}_{A,B}$ be the set of interpretations of A in B.

Corollary 8. If $A \leq B$, then $\text{Int}_{A,B}$ is Π_2^0 but not Σ_2^0.

Proof. Clearly $\text{Int}_{A,B}$ is Π_2^0. Suppose it is Σ_2^0. Evidently
$$A + \varphi \leq B \text{ iff } \exists t \in \text{Int}_{A,B}(B \vdash t(\varphi)).$$
It follows that $\{\varphi: A + \varphi \leq B\}$ is Σ_2^0, contradicting Theorem 12. ∎

In the next § we are going to prove that $\text{Int}_{A,B}$ is, in fact, a complete Π_2^0 set (Corollary 12).

§2. Faithful interpretability.

Let $t: S' \leq S$. t is a *faithful* interpretation of S' in S, $t: S' \trianglelefteq S$, if for every sentence φ, if $S \vdash t(\varphi)$, then $S' \vdash \varphi$. S' is *faithfully interpretable* in S, $S' \trianglelefteq S$, if there is a t such that $t: S' \trianglelefteq S$.

Most of the differences between $\leq$ and $\trianglelefteq$ are explained by the following lemma; for example, it is not true in general that if $S \dashv T$, then $S \trianglelefteq T$.

Lemma 6. If $Q \dashv S \trianglelefteq T$, then $T \dashv_{\Sigma_1} S$.

Proof. Suppose $t: S \trianglelefteq T$. Let σ be any Σ_1 sentence such that $T \vdash \sigma$. Clearly $t: Q + \neg\sigma \leq T + \neg t(\sigma)$. But then, by Lemma 1, $T + \neg t(\sigma) \vdash \neg\sigma$, and so $T \vdash t(\sigma)$. Since t is faithful, it follows that $S \vdash \sigma$. ∎

Our main aim in this § is to prove the following characterizations of $\trianglelefteq$.

§2. Faithful Interpretability

Theorem 13. $S \trianglelefteq T$ iff $S \leq T$ and for every φ, if $T \vdash \text{Pr}_\varnothing(\varphi)$, then $S \vdash \varphi$.

Theorem 14. $A \trianglelefteq B$ iff $A \dashv_{\Pi_1} B \dashv_{\Sigma_1} A$.

Corollary 9. (a) $S \trianglelefteq T$ iff for every k, $T \vdash \text{Con}_{S|k}$ and for every φ, if $T \vdash \text{Pr}_\varnothing(\varphi)$, then $S \vdash \varphi$.
 (b) If T is Σ_1-sound, then $S \trianglelefteq T$ iff $S \leq T$.
 (c) If $S \leq T \dashv S$, then $S \trianglelefteq T$.

Proof. (a) and (b) follow at once from Theorem 13 and Lemma 2. ♦
 (c) Suppose $T \vdash \text{Pr}_\varnothing(\varphi)$. Then, since T is essentially reflexive (Fact 11), $T \vdash \varphi$ and so, by assumption, $S \vdash \varphi$. Now use Theorem 13. ∎

By Corollary 9(c), Theorems 8, 9, 10 remain true when $\leq$ is replaced by $\trianglelefteq$.

Theorem 13 will be derived from the following two lemmas:

Lemma 7. Let $\sigma'(x)$ be a (Σ_1) formula binumerating S in T. There is then a (Σ_1) formula $\sigma(x)$ binumerating S in T and such that
(i) $\vdash \sigma(x) \to \sigma'(x)$, whence $\vdash \text{Con}_{\sigma'} \to \text{Con}_\sigma$,
(ii) for every sentence φ, if $T \vdash \text{Pr}_\sigma(\varphi)$, there is a q such that $T \vdash \text{Pr}_{S|q}(\varphi)$.

Lemma 8. Suppose $\sigma(x)$ numerates S in T and $T \vdash \text{Con}_\sigma$. There is then an interpretation $t: S \leq T$ such that for every φ, if $T \vdash t(\varphi)$, then $T \vdash \text{Pr}_\sigma(\varphi)$.

Proof of Lemma 7. Let $\sigma(x)$ be such that
$$PA \vdash \sigma(x) \leftrightarrow \sigma'(x) \wedge \forall yz \leq x(\text{Prf}_T(\text{Pr}_\sigma(\dot y),z) \to \text{Pr}_{\sigma'|z}(y)).$$
Then (i) is trivial.
 We now show that
(1)   if p is a proof of $\text{Pr}_\sigma(\varphi)$ in T, then $T \vdash \text{Pr}_{\sigma'|p}(\varphi)$.
Let p and φ be as assumed. Then, since $\varphi \leq p$,
$$T \vdash \neg \text{Pr}_{\sigma'|p}(\varphi) \to (\sigma(x) \to \sigma'(x) \wedge x \leq p).$$
It follows that
$$T \vdash \neg \text{Pr}_{\sigma'|p}(\varphi) \to (\text{Pr}_\sigma(\varphi) \to \text{Pr}_{\sigma'|p}(\varphi)).$$
But then, since $T \vdash \text{Pr}_\sigma(\varphi)$, we get $T \vdash \text{Pr}_{\sigma'|p}(\varphi)$, as desired.
 Since $\sigma'(x)$ binumerates S in T, it follows from (1) that (ii) holds.
 To show that $\sigma(x)$ binumerates S in T it suffices to show that for all φ and p,
$$T \vdash \text{Prf}_T(\text{Pr}_\sigma(\varphi),p) \to \text{Pr}_{\sigma'|p}(\varphi).$$
But this, too, follows at once from (1). ∎

Proof of Lemma 8. The following proof is a modification of the proof of

Theorem 4. The interpretation t constructed in that proof does not necessarily have the additional property that
(1) $T \vdash t(\varphi)$ implies $T \vdash \Pr_\sigma(\varphi)$.
To achieve this we proceed as follows. The function c, the set Z, and the formula $\zeta(x)$ are the same as before, but the definition of φ_n is different. Here we put
(2) $\varphi_n := \theta_n$ if $S + Z \vdash \bigwedge\{\varphi_m : m < n\} \to \theta_n$ or
$$(S + Z \nvdash \bigwedge\{\varphi_m : m < n\} \to \neg\theta_n \ \& \ n \in Y),$$
     $:= \neg\theta_n$ otherwise,
where Y is any set of natural numbers.

As before let $X = \{\varphi_n : n \in N\}$. Either θ_n or $\neg\theta_n$ is put in X. We put θ_n in X if putting $\neg\theta_n$ in X would make X inconsistent, and similarly for $\neg\theta_n$. Otherwise we put θ_n in X iff $n \in Y$. The idea is to achieve (1) by letting Y be formally represented by a sufficiently independent formula $\eta(x)$.

Let $\gamma(x) := \sigma(x) \vee \zeta(x)$. Let $\eta(x)$ be as in Theorem 2.10 with $\delta(x) := \Pr_\gamma(x)$. Next, as in the proof of Theorem 4, let $\chi(x,y)$ be the formalization of the result of turning the inductive definition of φ_n into an explicit definition using $\eta(x)$ to represent Y. Let $\xi(x) := \exists y \chi(x,y)$.

As in the proof of Theorem 4 we can now define an interpretation t of S in T such that
(3) $T \vdash t(\varphi) \leftrightarrow \xi(\varphi)$.
It remains to be shown that (1) holds.

Suppose $T \nvdash \Pr_\sigma(\varphi)$. We must then show that $T \nvdash t(\varphi)$. We have $T \nvdash \Pr_\gamma(\varphi)$ (see (6) in the proof of Theorem 4). For any $f \in 2^N$, let $Y_f = \{\Pr_\gamma(n)^{f(n)} : n \in N\}$. Now let $f(n)$ be such that $f(\varphi) = 1$ and
(4) $T + Y_f$ is consistent.
Next we define ψ_n as follows (compare (2)).
$$\psi_n := \theta_n \text{ if } \Pr_\gamma(\bigwedge\{\psi_m : m < n\} \to \theta_n) \in Y_f \text{ or}$$
$$(\Pr_\gamma(\bigwedge\{\psi_m : m<n\} \to \neg\theta_n) \notin Y_f \ \& \ \Pr_\gamma(\lambda_n) \notin Y_f),$$
     $:= \neg\theta_n$ otherwise,
where $\lambda_n := \bigwedge\{\psi_m : m < n\} \wedge \theta_n \to \varphi$. Let $g \in 2^N$ be such that
$$g(n) = 0 \text{ iff } \Pr_\gamma(\lambda_n) \notin Y_f$$
and set
$$Y_{f,g} = Y_f + \{\eta(n)^{g(n)} : n \in N\}.$$
Then, by (4) and the choice of $\eta(x)$,
(5) $T + Y_{f,g}$ is consistent.
Recalling the definition of $\chi(x,y)$, we can now show, by induction, that for every n, $T + Y_{f,g} \vdash \chi(\psi_n, n)$ and so
(6) $T + Y_{f,g} \vdash \xi(\psi_n)$.

§2. *Faithful Interpretability*   109

Next we show, by induction, that for every n,
(7) $Pr_\gamma(\wedge\{\psi_m:m<n\}\to\varphi) \notin Y_f$.
Note that, by (4), $\{\psi: Pr_\gamma(\psi) \in Y_f\}$ is closed under logical deduction. Since $Pr_\gamma(\varphi) \notin Y_f$, (7) holds for n = 0. Suppose (7) holds for n = k.

Case 1. $\psi_k := \theta_k$. Then either $Pr_\gamma(\wedge\{\psi_m:m<k\}\to\theta_k) \in Y_f$ or $Pr_\gamma(\wedge\{\psi_m:m<k+1\}\to\varphi) \notin Y_f$. In the latter case (7) holds for n = k+1. In the former case we have $Pr_\gamma(\wedge \{\psi_m:m<k\}\to\psi_k) \in Y_f$ and so (7) for n = k+1 follows from the inductive assumption.

Case 2. $\psi_k := \neg\theta_k$. Then
(8) $Pr_\gamma(\lambda_k) \in Y_f$.
For suppose $Pr_\gamma(\lambda_k) \notin Y_f$. If $Pr_\gamma(\wedge\{\psi_m:m<k\}\to\neg\theta_k) \in Y_f$, then $Pr_\gamma(\wedge\{\psi_m:m<k\}\wedge\theta_k\to\theta) \in Y_f$ for every θ and so, in particular, $Pr_\gamma(\lambda_k) \in Y_f$, contrary to assumption. So $Pr_\gamma(\wedge\{\psi_m:m<k\}\to\neg\theta_k) \notin Y_f$. But then $\psi_k := \theta_k$, a contradiction. This proves (8) and completes the proof of (7).

From (7) it follows that for some k, $\varphi := \neg\psi_k$. Hence, by (6), $T + Y_{f,g} \vdash \xi(\neg\varphi)$. But then, by (3) and (5), $T \nvdash t(\varphi)$. Thus, (1) holds and the proof is complete. ∎

Proof of Theorem 13. "If". By Corollary 2, there is a formula $\sigma'(x)$ binumerating S in T such that $T \vdash Con_{\sigma'}$. But then, by Lemma 7, there is a formula $\sigma(x)$ numerating S in T and such that $T \vdash Con_\sigma$ and Lemma 7(ii) holds. Now let t be as in Lemma 8. Then t: $S \leqslant T$. Let φ be any sentence of S such that $T \vdash t(\varphi)$. Then, by Lemma 8, $T \vdash Pr_\sigma(\varphi)$ and so there is a q such that $T \vdash Pr_{S|q}(\varphi)$. It follows that $T \vdash Pr_\varnothing(\wedge S|q\to\varphi)$ and so, by hypothesis, $S \vdash \wedge S|q \to \varphi$, whence $S \vdash \varphi$. Thus, t is faithful.

"Only if". Suppose $S \trianglelefteq T$. Then $S \leqslant T$. Let φ be such $T \vdash Pr_\varnothing(\varphi)$. Suppose t: $S \leqslant T$ is faithful. Let κ be the sentence saying that t is an interpretation of $\varnothing$ (logic) in T. Then, by Fact 12(b),
          $PA \vdash Pr_\varnothing(\varphi) \to Pr_\varnothing(\kappa\to t(\varphi))$.
But then $T \vdash Pr_\varnothing(\kappa\to t(\varphi))$. Since T is essentially reflexive, it follows that $T \vdash \kappa \to t(\varphi)$. But $T \vdash \kappa$ and so $T \vdash t(\varphi)$. But then, t being faithful, $S \vdash \varphi$, as desired. ∎

Proof of Theorem 14. Suppose first $A \dashv_{\Pi_1} B \dashv_{\Sigma_1} A$. Then, by Theorem 6, $A \leqslant B$. Suppose $B \vdash Pr_\varnothing(\varphi)$. Then, $Pr_\varnothing(\varphi)$ being Σ_1, it follows that $A \vdash Pr_\varnothing(\varphi)$. Since A is essentially reflexive, this implies that $A \vdash \varphi$. Hence, by Theorem 13, $A \trianglelefteq B$.

Next suppose $A \trianglelefteq B$. By Theorem 6, $A \dashv_{\Pi_1} B$, and, by Lemma 6, $B \dashv_{\Sigma_1} A$. ∎

The analogue of Theorem 11(a) for $\trianglelefteq$ now follows at once from Theorem 14 and Theorem 5.4(a) with, say, $\Gamma = \Pi_2$. We write $A \simeq B$ to mean that $A \trianglelefteq B \trianglelefteq A$.

Corollary 10. If $A \dashv B$, there is a sentence φ such that $A + \varphi \simeq B$.

The analogue of Theorem 11(b), on the other hand, is clearly false. (Let σ_k be Σ_1 sentences such that $T + \{\sigma_k: k < n\} \nvdash \sigma_n$ for every n and let $X = \{\sigma_k: k \in N\}$. Let σ be any Σ_1 sentence such that $T + \sigma \trianglelefteq T + X$. Then, by Lemma 6, $T + \sigma \vdash X$, whence $T + X \nvdash \sigma$ and so, again by Lemma 6, $T + X \ntrianglelefteq T + \sigma$.)

If S is finite, then $\{\varphi: S \leqslant T + \varphi\}$ is r.e., but if $\leqslant$ is replaced by $\trianglelefteq$ this is no longer true:

Corollary 11. Suppose $Q \dashv S \trianglelefteq T$. Then $X = \{\varphi: S \trianglelefteq T + \varphi\}$ is a complete Π_2^0 set.

Proof. By Theorem 13, X is Π_2^0. Let Y be any Π_2^0 set. By the proof of Theorem 5.6(a), for $\Gamma = \Sigma_1$, there is a formula $\xi(x)$ such that
(1)   if $k \in Y$, then $\xi(k)$ is Σ_1-conservative over T,
(2)   if $k \notin Y$, there is a Σ_1 sentence σ such that $T + \xi(k) \vdash \sigma$ and $S \nvdash \sigma$.
It is now sufficient to show that
(3) $Y = \{k: \xi(k) \in X\}$.
Suppose first $k \in Y$. Let ψ be any sentence such that $T + \xi(k) \vdash \Pr_\varnothing(\psi)$. Then, by (1), $T \vdash \Pr_\varnothing(\psi)$. But then, by Theorem 13, $S \vdash \psi$. Using Theorem 13 once again, we get $S \trianglelefteq T + \xi(k)$, i.e., $\xi(k) \in X$.

Next suppose $k \notin Y$. Let σ be as in (2). Since σ is Σ_1, $PA + \sigma \vdash \Pr_Q(\sigma)$ and so $PA + \sigma \vdash \Pr_\varnothing(\wedge Q \to \sigma)$. It follows that $T + \xi(k) \vdash \Pr_\varnothing(\wedge Q \to \sigma)$. On the other hand $S \nvdash \wedge Q \to \sigma$. Hence, by Theorem 13, $\xi(k) \notin X$. ∎

Finally, we improve Corollary 8 as follows.

Corollary 12. If $A \leqslant B$, then $\text{Int}_{A,B}$ is a complete Π_2^0 set.

Proof. Let $X = \{k: \forall m R(k,m)\}$, where $R(k,m)$ is r.e., be any Π_2^0 set. By Theorem 3.1, there is a formula $\rho(x,y)$ numerating $R(k,m)$ in B. Let $\alpha(x)$ be a formula binumerating A in B. Let $\sigma(x,y) :=$
$$\alpha(x) \wedge \text{Con}_{\alpha|x} \wedge \forall z \leqslant x \rho(y,z).$$
Then, by Lemma 3, for every k,
(1) $PA \vdash \text{Con}_{\sigma(x,k)}$.
By Lemma 2, for every n, $B \vdash \text{Con}_{A|n}$. It follows that
(2)   if $k \in X$, then $\sigma(x,k)$ binumerates A in B.
Also, clearly,
(3)   if $k \notin X$, there is an m such that $B \nvdash \exists x(m \leqslant x \wedge \sigma(x,k))$.

By (1) and the proof of Lemma 8, we can for each k, effectively find a translation t_k such that
(4) $t_k: \{\varphi: B \vdash \sigma(\varphi,k)\} \leqslant B$,
(5) if $B \vdash t_k(\varphi)$, then $B \vdash Pr_{\sigma(x,k)}(\varphi)$.
To complete the proof it suffices to show that
(6) $X = \{k: t_k \in Int_{A,B}\}$.
If $k \in X$, then, by (2) and (4), $t_k \in Int_{A,B}$. Suppose $k \notin X$. Let m be as in (3). Let θ be such that
$$PA \vdash \theta \leftrightarrow \neg Pr_{A|m}(\theta).$$
Then, since A is essentially reflexive,
(7) $A \vdash \theta$.
Since $A \leqslant B$, it follows, by Theorem 6, that $B \vdash \neg Pr_{A|m}(\theta)$. By the definition of $\sigma(x,y)$, this implies that
$$B \vdash Pr_{\sigma(x,k)}(\theta) \to \exists x(m \leqslant x \wedge \sigma(x,k)).$$
But then, by (3), $B \nvdash Pr_{\sigma(x,k)}(\theta)$ and so, by (5) and (7), $t_k \notin Int_{A,B}$. This proves (6) and so the proof is complete. ∎

Exercises for Chapter 6

In the following Exercises we assume that $PA \dashv T$ and that A, B, C are extensions of T.

1. (a) Show that there is a Π_1 sentence φ such that $Q + \varphi \nleqslant S$ and $Q + \neg\varphi \nleqslant S$ (compare Theorem 8.2).

   (b) Partially improve (a) as follows. Suppose $Q \dashv S$. Show that there is a (primitive) recursive sequence $\varphi_0, \varphi_1, \ldots$ of Π_1 sentences such that for every n and all i_k, $0 \leqslant i_k \leqslant 1$,
   $$Q + \varphi_0^{i_0} \vee \ldots \vee \varphi_n^{i_n} \nleqslant S + \varphi_0^{1-i_0} \wedge \ldots \wedge \varphi_n^{1-i_n}.$$
   [Hint: The set $\{\varphi: Q + \varphi \leqslant S + \neg\varphi\}$ is monoconsistent with Q.]

   (c) Partially improve (b) as follows. There is a finite subtheory PA_0 of PA such that Fact 10 holds with PA replaced by PA_0. Suppose $PA_0 \dashv S$. Show that there is a Γ formula $\gamma(x)$ such that for every Γ^d formula $\delta(x)$,
   $$PA_0 + \exists x(\delta(x) \wedge \gamma(x)) \vee \exists x(\neg\delta(x) \wedge \neg\gamma(x)) \nleqslant$$
   $$S + \forall x(\delta(x) \to \neg\gamma(x)) \wedge \forall x(\neg\delta(x) \to \gamma(x)).$$
   [Hint: Use Exercise 2.22(c).]

2. (a) Suppose $A \dashv B \nleqslant A$. Show that there is a C such that $A \dashv C \dashv B$ and $B \nleqslant C \nleqslant A$. [Hint: Use Orey's compactness theorem.]

   (b) Suppose $A < B$. Show that there is a C such that $A < C < B$.

3. The proof of Theorem 4 actually yields the following stronger result: There is a finite subtheory PA_σ of PA such that $S \leqslant PA_\sigma + \{\sigma(\varphi): \varphi \in S\} + Con_\sigma$. Use this to prove the following:

(a) T is interpretable in a bounded subtheory of T (compare Corollary 4.1(a) and Theorem 3).

(b) If $\tau(x)$ numerates T in a finite subtheory of T, then $T + Con_\tau \not\leqslant T$ (compare Theorem 2 and Corollary 2.4). (This can also be proved by combining Theorem 2(b) and the proof of Corollary 2.4.)

4. (a) Suppose σ_0, σ_1 are Σ_1 sentences such that $T + \sigma_i \leqslant T$, $i = 0, 1$. Show that $T + \sigma_0 \wedge \sigma_1 \leqslant T$.

(b) Show that there is a Π_1^0 set X of Σ_1 sentences such that $T + X \not\leqslant T$ and $T + Y \leqslant T$ for every finite (and so for every r.e.) subset Y of X (compare Theorem 5). [Hint: Let $\tau(x)$ be a PR binumeration of T. Let $\rho(x,y)$ be a PR formula such that $Z = \{k: \exists m PA \vdash \rho(k,m)\}$ is not recursive. Let $\gamma(x,y) := \tau(x) \wedge \forall z \leqslant x \neg \rho(y,z)$. Finally, let $X = \{\neg Con_{\gamma(x,k)}: k \notin Z\}$.]

5. Improve Corollary 3 by showing that $Int_{A,B} \subseteq Int_{A+\sigma, B+\sigma}$.

6. (a) Use Exercise 2.17(b) to give an alternative proof of Theorem 8.

(b) Use Exercise 2.18 and Theorem 8 to give another proof of Theorem 9.

7. Prove the following strong partial converse of Theorem 1: Suppose for every PR binumeration $\tau(x)$ of T, there is a formula $\sigma(x)$ numerating S in T such that $T \vdash Con_\tau \to Con_\sigma$. Then $S \leqslant T$. [Hint: Use Lemma 2 and Exercise 2.18.]

8. Let $\tau(x)$ be a PR binumeration of T. Show that there is a (Π_1, Σ_1) sentence φ such that $\tau + \varphi \leqslant \tau$ and $T + \varphi \not\leqslant T$ (cf. Exercise 2.27).

9. Let θ be a Π_1 Rosser sentence for T and let $\psi :=$
$$\forall u(Prf_T(\neg\theta, u) \to \exists z \leqslant u Prf_T(\theta, z)).$$
Show that $T + \theta \equiv T + \neg\psi$, $T + \psi \equiv T + \neg\theta$, $T + \theta < T + Con_T$, $T + \psi < T + Con_T$.

10. Suppose X is r.e. and monoconsistent with T. Let $\rho(x,y)$ be a PR formula such that $X = \{k: \exists m PA \vdash \rho(k,m)\}$.

(a) Let φ be such that
$$PA \vdash \varphi \leftrightarrow \forall z(Con_{T|z+\varphi} \to \neg\rho(\varphi, z)).$$

Show that $T + \varphi \leqslant T$ and $\varphi \notin X$.

(b) The sentence φ in (a) is Π_2. This can be improved. Let χ be such that
$$PA \vdash \chi \leftrightarrow \exists z(\neg Con_{T|z+\chi} \wedge \forall u \leqslant z \neg \rho(\chi,u)).$$
Then χ is Σ_1. Show that $T + \chi \leqslant T$ and $\chi \notin X$ (compare Theorem 9).

11. (a) Let φ be such that
$$PA \vdash \varphi \leftrightarrow \forall z(Con_{T|z+\varphi} \to Con_{T|z+\neg\varphi}).$$
Show that φ is an Orey sentence for T (compare Theorem 10).

(b) Suppose $A \dashv B$. Let φ be such that
$$PA \vdash \varphi \leftrightarrow \forall z(Con_{A|z+\varphi} \to Con_{B|z}).$$
Show that $A + \varphi \equiv B$ (compare Theorem 11(a)).

(c) Let φ be such that
$$PA \vdash \varphi \leftrightarrow \forall z(Con_{A|z+\varphi} \to Con_{B|z+\neg\varphi}).$$
Show that $A + \varphi \equiv B + \neg\varphi$.

12. (a) Show that
$$PA \vdash \forall x(Con_{S_0|x} \to Con_{S_1|x}) \leftrightarrow (Con_{S_1} \vee \exists x(\neg Con_{S_0|x} \wedge \forall y < x Con_{S_1|x})).$$
Conclude that the sentences φ of Exercise 11 are Δ_2 (compare Exercise 5.6(a) and Theorem 7.8). In particular, there is a Δ_2 Orey sentence for T.

(b) Show that no Orey sentence for T is B_1.

13. Let $\tau^*(x)$ be as in Theorem 2.7. In Theorem 4 let $\sigma(x) := \tau^*(x)$ and $S = T$. Next let $\xi(x)$ be as in (14) of the proof of Theorem 4. Let φ be such that $PA \vdash \varphi \leftrightarrow \neg\xi(\varphi)$. Show that φ is an Orey sentence for T.

14. Let $\tau(x)$ be any formula binumerating T in T. Let φ be such that
$$PA \vdash \varphi \leftrightarrow \neg Pr_\tau(\varphi).$$
Show that
(i) $T + \neg\varphi \leqslant T$,
(ii) for every n, $T + \varphi \leqslant T + Pr_{T|n}(Con_\tau)$.
Let $\tau(x)$ be the formula $\tau^*(x)$ mentioned in Theorem 2.7. Conclude that φ is then an Orey sentence for T.

15. (a) Suppose $T \leqslant S$. Show that for every Π_2^0 set X, there is a Σ_1 formula $\xi(x)$ such that
$$\{k: T + \xi(k) \leqslant S\}$$
(compare Theorem 12). [Hint: Let R(k,m) be an r.e. relation such that $X = \{k: \forall m R(k,m)\}$. There is a Σ_1 formula $\rho(x,y)$ such that

if $R(k,m)$, then $Q \vdash \rho(k,m)$,
if not $R(k,m)$, then $Q + \rho(k,m) \not\trianglelefteq S$
(Lemma 3.1). Let $\xi(x)$ be such that
$$PA \vdash \xi(k) \leftrightarrow \exists z(\neg Con_{T|z+\xi(k)} \wedge \forall u \leqslant z \rho(k,u));$$
compare Exercise 10(b).]

(b) Let $\delta(x)$ and S be such that $T \vdash \delta(x) \wedge \delta(y) \to x = y$ and for every k, $T + \delta(k) \leqslant S$. Show that for every Π_2^0 set X, there is a sentence φ such that
$$X = \{k\colon T + \varphi \wedge \delta(k) \leqslant S\}.$$
[Hint: Let $\rho(x,y)$ be as in (a) and let φ be such that
$$T \vdash \varphi \leftrightarrow \exists u(\delta(u) \wedge \exists y(\neg[\Pi_1](\delta(\dot{u}) \wedge \varphi, y) \wedge \forall z \leqslant y \rho(u,z))).]$$

(c) Suppose $T \leqslant S$. Show that there is a formula $\delta(x)$ as in (b). [Hint: Use Exercise 5.7(b).]

16. (a) By Orey's compactness theorem (Theorem 5), there is a function f such that for every sentence φ, if $T + \varphi \not\trianglelefteq T$, then $T|f(\varphi) + \varphi \not\trianglelefteq T$. Show that $f(k)$ cannot be recursive.

(b) By Theorem 6, there is a function g such that for every sentence φ, if $T + \varphi \not\trianglelefteq T$, then $g(\varphi)$ is a Π_1 sentence such that $T + \varphi \vdash g(\varphi)$ and $T \not\vdash g(\varphi)$. Show that $g(k)$ cannot be recursive.

17. Show that, even if T is not Σ_1-sound, there is a Σ_1 formula $\tau(x)$ binumerating T in T such that $Pr_\tau(x)$ numerates $Th(T)$ in T (by Exercise 2.25, $\tau(x)$ cannot be PR).

18. Show that if $A \leqslant B$, then $\{\varphi \in \Sigma_1 \colon A + \varphi \trianglelefteq B\}$ is a complete Π_2^0 set.

19. (a) Show that if S is finite and $Q \dashv S \trianglelefteq T$, then the set of faithful interpretations of S in T is a complete Π_2^0 set. [Hint: First show that there is a sentence θ such that $S \not\vdash \theta$ and $S + \theta \leqslant T$.]

(b) Suppose $A \trianglelefteq B$. Show that the set of faithful interpretations of A in B is a complete Π_2^0 set.

20. S is X-*faithfully interpretable* in S', $S \trianglelefteq_X S'$, if there is an interpretation t: S $\leqslant$ S' which is X-*faithful* in the sense that for every $\varphi \in X$, if $S' \vdash t(\varphi)$, then $S \vdash \varphi$. Show that

(i) $S \trianglelefteq_X T$ iff $S \leqslant T$ and for every $\varphi \in X$, if there is an m such that $T \vdash Pr_{S|m}(\varphi)$, then $S \vdash \varphi$,

(ii) if $S \leqslant T$, then $S \trianglelefteq_X T$, where $X = \{\varphi \colon S \trianglelefteq_{\{\varphi\}} T\}$,

(iii) if $S \trianglelefteq_X T$ and $S \leqslant T' \dashv T$, then $S \trianglelefteq_X T'$,

(iv) if $S \trianglelefteq T$ and $S \dashv S' \leq T' \dashv T$, then $S' \trianglelefteq T'$,
(v) $\trianglelefteq$ cannot be replaced by $\trianglelefteq_X$ in (iv),
(vi) $A \trianglelefteq_X B$ iff $A + (\text{Th}(B) \cap \Sigma_1) \dashv_X A \dashv_{\Pi_1} B$,
(vii) $A \trianglelefteq B$ iff $A \trianglelefteq_{\Sigma_1} B$ iff $A \trianglelefteq_{\{\varphi\}} B$ for every (Σ_1) sentence φ.

21. S is *cointerpretable in* T, S co-$\leq$ T, if there is a *cointerpretation* of S in T, i.e., an interpretation in S such that for every φ, if $S \vdash t(\varphi)$, then $T \vdash \varphi$.
    (a) Show that the following conditions are equivalent.
    (i) S co-$\leq$ T.
    (ii) For every φ, if $S \vdash \text{Pr}_\emptyset(\varphi)$, then $T \vdash \varphi$.
    Conclude that $S \trianglelefteq T$ iff $S \leq T$ and T co-$\leq$ S.
    (b) Show that A co-$\leq$ B iff $A \dashv_{\Sigma_1} B$.

22. Show that $A \dashv_{\Pi_n} B$ iff there is a t: $A \leq B$ such that for every Π_n sentence ψ, $B \vdash t(\psi) \to \psi$ (compare Theorem 6; note that for every t: $A \leq B$ and every Π_1 sentence ψ, $B \vdash t(\psi) \to \psi$, by Lemma 1). [Hint: "Only if". For every k and every Π_n sentence φ, $B \vdash \text{Pr}_{A|k}(\varphi) \to \varphi$. Use this to construct a formula $\alpha(x)$ binumerating A in B and such that $\text{PA} \vdash \text{Con}_\alpha$ and $B \vdash \chi \to \alpha(\chi)$ for every Σ_n sentence χ.]

Notes for Chapter 6
The general concept (*relative*) *interpretation* due to Tarski (cf. Tarski, Mostowski, Robinson (1953); in keeping with recent usage we omit "relative"); it is an important tool in proofs of (relative) consistency and (un)decidability. The investigation of interpretability for its own sake was initiated by Feferman (1960) (see also Feferman (1997)). Theorems 1, 2, 3 are due to Feferman (1960); concerning the (im)possibility of improving Theorem 3, see Exercise 3(b). Theorem 4 is due to Feferman (1960) building on work of Bernays (Hilbert and Bernays (1939)) and Wang (1951); for a strengthening of Theorem 4, see Exercise 3. Lemma 2 is implicit in Feferman (1960), all but explicit in Orey (1961), and fully explicit in Hájek (1971). Similar, but more complicated, characterizations of $\leq$ for other pairs of theories have been obtained by Friedman (cf. Smoryński (1985a), Visser (1990)), Pudlák (1985) (cf. also Hájek and Pudlák (1993)), and Shavrukov (1997). Corollary 2 is due to Orey (1961). Theorem 5 is due to Orey (1961) (cf. also Feferman (1960)); for a general result in the opposite direction, see Pudlák (1985). Theorem 6 was first stated by Guaspari (1979) and Lindström (1979); for a more general result, due to

Guaspari (1979), see Exercise 22. Corollary 4 is due to Goryachev (1986). Theorem 8 is due to Feferman (1960) (with a different proof; see Exercise 6(a)). Lemma 4 is due to Švejdar (1978). Theorem 9 (with a different proof, see Exercise 10(a)) and Corollaries 5(a) and 6 are essentially due to Hájek (1971) (cf. also Hájková and Hájek (1972)); for yet another proof, see Exercise 6(b). Corollary 5(b) has also been pointed out by Guaspari (1979); for a related result, see Exercise 8. Corollary 5 answers a question left open in Feferman (1960). Theorem 10 less the references to the set X is due to Orey (1961); the full result is proved in Lindström (1979), (1984a); a generalization of Orey's theorem follows at once from Theorem 6, Theorem 5.7, and Exercise 5.13; for an even more general result, see Strannegård (1997), (1999); related results, for certain nonreflexive theories, requiring methods not explained here, can be found in Hájek and Pudlák (1993). For more information on Orey sentences, see Exercises 11(a), 12(b), 13, 14. The result on finite conservative extensions mentioned just before Lemma 5 is due to Kleene (1952b) (cf. also Kaye (1991)). Theorem 11 is due to Lindström (1979) (see Exercise 11(b)) and (1984a); by Exercises 11(b) and 12, the sentence φ in Theorem 11(a) can be taken to be Δ_2 (cf. also Theorem 7.8). Theorem 12 is essentially due to Solovay (cf. Hájek (1979)) (with a different proof); the present proof is from Lindström (1984a) (see also Exercise 15).

The concept *faithful interpretation* was introduced in Feferman, Kreisel, Orey (1960). They observed that if $Q \dashv S \trianglelefteq S'$ and S is Σ_1-sound, so is S' (see Lemma 6). Theorems 13 and 14 are due to Lindström (1984c); see also Exercise 20. Corollary 9(b) is due to Feferman, Kreisel, Orey (1960). Lemma 7 is due to Lindström (1984c); the present proof is an instance of a general argument described in Lindström (1988). Lemma 8 is due to Lindström (1984c), but the main idea of the proof, to introduce the set Y and represent Y by a sufficiently independent formula, is taken from Feferman, Kreisel, Orey (1960). Corollary 10 is due to Lindström (1984c); Exercise 7.8, below, is an improvement of Corollary 10. Corollaries 11, 12 are due to Lindström (1984c); for related results, see Exercises 18 and 19.

An alternative notion of interpretability, *feasible interpretability*, has been studied by Verbrugge (1992), (1994). For any formal entity q, formula, proof, etc., let |q| be the length of q, i.e., the number of (instances of) symbols occurring in q. S is *feasibly interpretable* in T, $S \leqslant_f T$, if there is an interpretation t: $S \leqslant T$ which is *feasible* in the sense that there is a polynomial P(n) such that for every $\varphi \in S$, there is a proof p of t(φ) in T such that $|p| \leqslant P(|\varphi|)$. Clearly, $\{\varphi: S + \varphi \leqslant_f T\}$ is Σ_2^0; in fact, it is complete Σ_2^0. Thus, by

Theorem 12, $S \leqslant T$ does not imply $S \leqslant_f T$ (cf. Verbrugge (1994)).

Exercise 1(a) and (b) are (essentially) due to Montague (1957), (1962). Exercise 2(a) is due to Jeroslow (1971a); Exercise 2(b) is due to Švejdar (1978). Exercise 3 is due to Feferman (1960). Exercise 4(b) (with a different proof) is essentially due to Orey (1961). Exercise 9 is due to Švejdar (1978). Exercise 10(a) is essentially due to Hájek (1971), see also Hájková and Hájek (1972). Exercise 11(a) is due to Lindström (1979) and Švejdar (1978). Exercise 12 was pointed out to me by Franco Montagna (compare Theorem 7.8). Exercise 13 is due to Orey (1961). Exercise 16(a) is due to Jeroslow (1971b). Exercise 21 is due to Dzhaparidze (1993); see also Japaridze and de Jongh (1998). Exercise 22 is due to Guaspari (1979).

7. DEGREES OF INTERPRETABILITY

Suppose PA ⊣ T. We shall use A, B, etc. for extensions of T. (Thus, T, A, B, etc. are essentially reflexive.) The relation ⩽ of interpretability is reflexive and transitive. Thus, the relation ≡ of mutual interpretability (restricted to extensions of T) is an equivalence relation; its equivalence classes will be called *degrees* (*of interpretability*) and will be written a, b, c, etc. $\mathbf{D}_T$ is the set of degrees of extensions of T. A is of degree a if A ∈ a and d(A) is the degree of A. The relation ⩽ among degrees is the relation induced by the relation ⩽ among theories: d(A) ⩽ d(B) iff A ⩽ B. $\mathbf{D}_T = (D_T, ⩽)$, the partially ordered set of degrees defined in this way, will be studied in some detail in this chapter.

§1. **Algebraic properties.** In this § we restrict ourselves to purely algebraic properties of $\mathbf{D}_T$. First we define the theory A^T and the operations ↓ and ↑ on theories as follows.
$$A^T = T + \{Con_{A|k}: k \in N\},$$
$$A{\downarrow}B = T + \{Con_{A|k} \vee Con_{B|k}: k \in N\},$$
$$A{\uparrow}B = T + \{Con_{A|k} \wedge Con_{B|k}: k \in N\}.$$
From Lemma 6.2 and Theorem 6.6, we get the following:

Lemma 1. (a) A ⩽ B iff A^T ⊣ B. Thus, A^T ≡ A and A ⩽ B iff A^T ⊣ B^T.
 (b) A ⩽ B, C iff A ⩽ B↓C.
 (c) A, B ⩽ C iff A↑B ⩽ C.

The following lemma is little more than a restatement of Lemma 4.4.

Lemma 2. If θ is Π_1 and A ⊢ θ, there is a k such that PA ⊢ $Con_{A|k}$ → θ.

Instead of A↓B it is sometimes convenient to use the theory A∨B defined by
$$A \vee B = \{\varphi \vee \psi: \varphi \in A \ \& \ \psi \in B\}.$$
Th(A∨B) = Th(A) ∩ Th(B). Evidently, A↓B ⊣ A∨B and, by Lemma 2, A∨B $\dashv_{\Pi_1}$ A↓B. But then, by Theorem 6.6, that A∨B ⩽ A↓B and so A∨B ≡ A↓B. It follows that for every sentence φ, (A + φ)↓(A + ¬φ) ⩽ A.

From Lemma 2 and Lemma 6.1 we get:

§1. Algebraic Properties

Lemma 3. For every Π_1 sentence π, $T + \pi \leq A \!\uparrow\! B$ iff $A \!\uparrow\! B \vdash \pi$ iff there are Π_1 sentences φ, ψ such that $A \vdash \varphi$, $B \vdash \psi$, and $T + \varphi \wedge \psi \vdash \pi$.

For $A \in a$ and $B \in b$, let $a \cap b = d(A \!\downarrow\! B)$ and $a \cup b = d(A \!\uparrow\! B)$. By Lemma 1, $\cap$ and $\cup$ are well-defined, $a \cap b$ is the g.l.b. of a and b and $a \cup b$ is l.u.b. of a and b. Thus, we have proved part of the following:

Theorem 1. $\mathbf{D}_T$ is a distributive lattice.

To prove distributivity we need the following lemma whose proof is left to the reader.

Lemma 4. (a) For every k, there is an m such that
$$PA \vdash Con_{(A \vee B)|m} \to Con_{A|k} \vee Con_{B|k}.$$
(b) For every k, there is an m such that
$$PA \vdash Con_{A|m} \vee Con_{B|m} \to Con_{(A \vee B)|k}.$$

Proof of Theorem 1. Let $D = A^T \vee (B \!\uparrow\! C)$ and $E = (A \vee B) \!\uparrow\! (A \vee C)$. To prove that $\mathbf{D}_T$ is distributive, it is, by Lemma 1, sufficient to show that $D \dashv\vdash E$.

Let k be arbitrary. By Lemma 4(a), there is an m such that
$$PA \vdash Con_{(A \vee B)|m} \to Con_{A|k} \vee Con_{B|k},$$
$$PA \vdash Con_{(A \vee C)|m} \to Con_{A|k} \vee Con_{C|k}.$$
But then
$$E \vdash Con_{A|k} \vee (Con_{B|k} \wedge Con_{C|k}).$$
It follows that $D \dashv E$. The proof that $D \vdash E$ is similar. ∎

$\mathbf{D}_T$ has a minimal element $0_T = d(T)$ and a maximal element 1_T, the common degree of all inconsistent theories.

In our next result we answer a number of standard questions concerning $\mathbf{D}_T$; in particular, it follows that $\mathbf{D}_T$ is dense.

Theorem 2. Suppose $a < b < 1_T$, $d_0 \not\leq a$, and $b \not\leq d_1$. There are then degrees c_0, c_1 such that $a < c_i < b$, $d_0 \not\leq c_i \not\leq d_1$, $i = 0, 1$, $c_0 \cap c_1 = a$, and $c_0 \cup c_1 = b$.

We derive this from:

Lemma 5. Suppose X is r.e. and monoconsistent with PA. Let θ be any true Π_1 sentence. There are then Π_1 sentences θ_0, θ_1 such that
(i) $PA \vdash \theta_0 \vee \theta_1$,
(ii) $PA \vdash \theta_0 \wedge \theta_1 \to \theta$,

(iii) $\theta_i^j \notin X$, $i, j = 0, 1$.

Proof. We may assume that if $\varphi \in X$ and $PA \vdash \varphi \to \psi$, then $\psi \in X$. Let $\theta := \forall x \gamma(x)$, where $\gamma(x)$ is PR. Let $R(k,m)$ be a primitive recursive relation such that $X = \{k: \exists m R(k,m)\}$ and let $\rho(x,y)$ be a PR binumeration of $R(k,m)$. Finally, let θ_0 and θ_1 be such that
(1) $\quad PA \vdash \theta_0 \leftrightarrow \forall y((\rho(\theta_0,y) \vee \neg\gamma(y)) \to \exists z \leqslant y \rho(\theta_1,z))$,
(2) $\quad PA \vdash \theta_1 \leftrightarrow \forall z(\rho(\theta_1,z) \to \exists y < z(\rho(\theta_0,y) \vee \neg\gamma(y)))$.
Then (i) and (ii) follow directly (cf. Lemma 1.3).

Suppose $\theta_0 \in X$ or $\theta_1 \in X$ and let m be the least number such that $R(\theta_0,m)$ or $R(\theta_1,m)$. If $R(\theta_1,m)$, then $\theta_1 \in X$. Also, by (2), and since θ is true, $PA \vdash \neg\theta_1$, a contradiction. It follows that not $R(\theta_1,m)$ and, therefore, $R(\theta_0,m)$. But then $\theta_0 \in X$ and, by (1), $PA \vdash \neg\theta_0$, again a contradiction. Thus, $\theta_0 \notin X$ and $\theta_1 \notin X$.

Finally, if $\neg\theta_i \in X$, by (i), $\theta_{1-i} \in X$. It follows that $\neg\theta_0 \notin X$ and $\neg\theta_1 \notin X$. ∎

Proof of Theorem 2. Let $A \in a$, $B \in b$, $D_i \in d_i$. By Orey's compactness theorem (Theorem 6.5) there are sentences ψ, χ such that $B \vdash \psi$, $\psi \not\leqslant A$, $D_0 \vdash \chi$, $\chi \not\leqslant A$. By Theorem 6.6, there is a Π_1 sentence π such that $B \vdash \pi$, $A \nvdash \pi$, and $D_1 \nvdash \pi$. Let

$$X_0 = \{\varphi: A \vdash \varphi \vee \pi\}, \qquad X_1 = \{\varphi: \psi \leqslant A + \neg\varphi\},$$
$$X_2 = \{\varphi: \chi \leqslant A + \neg\varphi\}, \qquad X_3 = \{\varphi: D_1 \vdash \varphi \vee \pi\}.$$

Let $X = X_0 \cup X_1 \cup X_2 \cup X_3$. Then X is r.e. (cf. Lemma 6.5). It is also easy to verify that X is monoconsistent with PA. By Lemma 5, there are then Π_1 sentences θ_0, θ_1 such that
(1) $\quad PA \vdash \theta_0 \vee \theta_1$,
(2) $\quad PA \vdash \theta_0 \wedge \theta_1 \to Con_B$,
(3) $\quad \theta_i^j \notin X$, $i, j = 0, 1$.
Let $e_i = d(A+\theta_i)$, $i = 0, 1$. Then $a \leqslant e_i$, $b \not\leqslant e_i$, since $\neg\theta_i \notin X_1$. $d_0 \not\leqslant e_i$, since $\neg\theta_i \notin X_2$. Let $c_i = e_i \cap b$. Then $c_i < b$ and $d_0 \not\leqslant c_i$. If $c_i \leqslant a$, then, since θ_i is Π_1, $A \vdash \theta_i \vee \pi$, contradicting the fact that $\theta_i \notin X_0$. Thus, $c_i \not\leqslant a$ and so $a < c_i$. Similarly, $c_i \not\leqslant d_1$, since $\theta_i \notin X_3$. By (1), $c_0 \cap c_1 = a$. By (2), Theorem 6.4, and Lemma 3, $e_0 \cup e_1 \geqslant b$, whence $c_0 \cup c_1 = b \cap (e_0 \cup e_1) = b$. ∎

From Lemma 5 we can also derive the following:

Corollary 1. T is not Σ_1-sound iff there are degrees $a_0, a_1 < 1_T$ such that $a_0 \cup a_1 = 1_T$ (and $a_0 \cap a_1 = 0_T$).

Proof. Suppose T is Σ_1-sound. Let $a, b < 1_T$, $A \in a$, $B \in b$. Then $A \hat{\uparrow} B$ is consistent and so $a \cup b < 1_T$. Next, suppose T is not Σ_1-sound. There is then a

§1. *Algebraic Properties*

true Π_1 sentence θ such that $T \vdash \neg\theta$. Let θ_i be as in Lemma 5 with $X = \text{Th}(T)$. Let $a_i = d(T+\theta_i)$. Then $a_i < 1_T$, by Lemma 5(iii), and $a_0 \cap a_1 = 0_T$, by Lemma 5(i). Finally, by Lemma 3, $(T + \theta_0) \uparrow (T + \theta_1) \vdash \theta$. Since $T \vdash \neg\theta$, it follows that $(T + \theta_0) \uparrow (T + \theta_1)$ is inconsistent and so $a_0 \cup a_1 = 1_T$. ∎

By Corollary 1, if PA ⊣ S and S is Σ_1-sound but T is not, then $\mathbf{D}_S$ and $\mathbf{D}_T$ are not isomorphic. But suppose S and T are both Σ_1-sound. It is an open problem if this implies that $\mathbf{D}_S$ and $\mathbf{D}_T$ are isomorphic.

Given that there are $c_0, c_1 >$ a such that $c_0 \cap c_1 = a$, we may ask if any b such that $a < b < 1_T$ *caps to* a in the sense that there is a $c >$ a such that $b \cap c = a$. (Dually, b *cups to* a if there is a $c <$ a such that $b \cup c = a$.) In our next result this question and its dual are answered in the negative. We write a $\ll_\cap$ b to mean that $a < b$ and b does not cap to a. Dually, $a \ll_\cup b$ means that $a < b$ and a does not cup to b.

Theorem 3. (a) Suppose $0_T < a \not\leq c$. There is a b such that $0_T < b \ll_\cup a$ and $b \not\leq c$.

(b) Suppose $c \not\leq a < 1_T$. There is a b such that $a \ll_\cap b < 1_T$ and $c \not\leq b$.

Proof. (a) Let $A \in a$ and $C \in c$. There is a Π_1 sentence θ such that $A \vdash \theta$ and $C \not\vdash \theta$. Let $X = \text{Th}(C + \neg\theta)$. X is r.e. and (mono)consistent with $T + \neg\theta$. By Theorem 5.2, there is a Π_1 sentence $\psi \notin X$ such that ψ is Σ_1-conservative over $T + \neg\theta$. Let $B = T + \psi \vee \theta$ and $b = d(B)$. Then $0_T < b \not\leq c$ and $b \leq a$. Suppose $b \cup d = a$. Let $D \in d$. Then, by Lemma 6.2, there is an m such that $T + \psi + \text{Con}_{D|m} \vdash \theta$ and so $T + \neg\theta + \psi \vdash \neg\text{Con}_{D|m}$. Since ψ is Σ_1-conservative over $T + \neg\theta$, it follows that $T + \neg\theta \vdash \neg\text{Con}_{D|m}$ and so $D \vdash \theta$. Thus, $d \geq b$ and so $d = b \cup d = a$. ◆

The proof of the following from Lemma 6.2, Theorem 6.6, and Lemma 2 is straightforward.

Lemma 6. The following conditions are equivalent:
(i) $A \downarrow B \leq C$.
(ii) $A \leq C + \neg\text{Con}_{B|k}$ for every k.
(iii) $A \leq C + \neg\theta$ for every Π_1 sentence θ such that $B \vdash \theta$.

Let σ be any Σ_1 sentence. By Corollary 6.3, the degree $d(A+\sigma)$ is uniquely determined by σ and $d(A)$. Thus, we may denote the former by $d(A) + \sigma$. If X is an r.e. set of Σ_1 sentences, then, by Theorem 6.11(b), there is a Σ_1 sentence σ such that $d(A+X) = d(A) + \sigma$.

Lemma 7. The following conditions are equivalent:
(i) $a \ll_\cap b$.
(ii) $a < b$ and for every Σ_1 sentence σ, if $b \leq a + \sigma$, then $a + \sigma = 1_T$.

Proof. Suppose (i) holds. Let $A \in a$ and $B \in b$. Let σ be Σ_1 and such that $b \leq a + \sigma$. Then $B \downarrow (A + \neg\sigma) \leq (A + \sigma) \downarrow (A + \neg\sigma) \leq A$. Hence, by assumption, $A + \neg\sigma \leq A$, whence $A \vdash \neg\sigma$ and so $a + \sigma = 1_T$. Thus, (ii) holds.

Next suppose (ii) holds. Let c be such that $b \cap c = a$. Let $A \in a$, $B \in b$, $C \in c$. Let θ be any Π_1 sentence provable in C. It suffices to show that $A \vdash \theta$. By Lemma 6, $B \leq A + \neg\theta$. But then, by assumption, $A \vdash \theta$, as desired. ∎

Lemma 8. If π is Π_1, $A \leq B + \pi$, and $\neg\pi$ is Π_1-conservative over A, then $d(A) \ll_\cap d(B + \pi)$.

Proof. Suppose $B + \pi \leq A + \sigma$. Then, by Lemma 6.1, $A + \sigma \vdash \pi$, whence $A + \neg\pi \vdash \neg\sigma$ and so $A \vdash \neg\sigma$, in other words, $A + \sigma$ is inconsistent. Now use Lemma 7. ∎

Proof of Theorem 3(b). Let $A \in a$, $C \in c$. By Theorem 6.5, there is a sentence ψ such that $C \vdash \psi \not\leq A$. Let $X = \{\varphi: \psi \leq A + \neg\varphi\}$. Then, by Theorem 5.2, there is a Σ_1 sentence $\chi \notin X$ such that χ is Π_1-conservative over A. Let $B = A + \neg\chi$ and $b = d(B)$. Then $c \not\leq b < 1_T$. Finally, by Lemma 8, $a \ll_\cap b$. ∎

From Theorem 6.4 and Lemma 8, and Theorem 5.1, we get the following (compare Theorem 6.2):

Corollary 2. $d(A) \ll_\cap d(T + \text{Con}_A)$.

Theorem 3(a) leads to the problem if for any $a < 1_T$, there is a b such that $a \ll_\cup b < 1_T$. (The dual of this is false: if $0_T < b < a$ and not $0_T \ll_\cap a$, then not $b \ll_\cap a$.) We now show that the answer is negative.

a is a *cupping* degree if $a < 1_T$ and a cups to every b such that $a \leq b < 1_T$. Let
$$\text{CON}_T = \{a < 1_T: a = d(T + \text{Con}_\tau) \text{ for some PR binumeration } \tau(x) \text{ of } T\}.$$
By Corollary 2.5, $\text{CON}_T \neq \emptyset$.

Theorem 4. Every member of CON_T is a cupping degree.

Proof. Suppose $a = d(T + \text{Con}_\tau) < 1_T$, where $\tau(x)$ is a PR binumeration of T. Let b be any degree such that $a \leq b < 1_T$. Let $B \in b$. We want to define a degree d such that $d \not\geq a$ and $a \cup d \geq b$. The obvious way to try is to let $d =$

§1. Algebraic Properties

$d(T+\theta)$, where
$$\theta := \forall u(\mathrm{Prf}_B(\bot,u) \to \exists z<u \mathrm{Prf}_\tau(\bot,z)).$$
But it seems difficult to prove, and may not even be true, that $d \geqslant a$ and so we have to proceed in a somewhat different way.

Let φ be such that
$$\mathrm{PA} \vdash \varphi \leftrightarrow \forall z(\mathrm{Prf}_\tau(\varphi,z) \to \exists u \leqslant z \mathrm{Prf}_B(\bot,u)),$$
and let
$$\psi := \forall u(\mathrm{Prf}_B(\bot,u) \to \exists z<u \mathrm{Prf}_\tau(\varphi,z)).$$
Then
(1) $\quad T \nvdash \varphi$,
(2) $\quad \mathrm{PA} \vdash \varphi \vee \psi$,
(3) $\quad \mathrm{PA} \vdash \varphi \wedge \psi \to \mathrm{Con}_B$.

Clearly, $\mathrm{PA} \vdash \neg\varphi \to \mathrm{Pr}_\tau(\varphi)$. Since $\neg\varphi$ is Σ_1, we also have, by provable Σ_1-completeness, $\mathrm{PA} \vdash \neg\varphi \to \mathrm{Pr}_\tau(\neg\varphi)$. Thus,
(4) $\quad \mathrm{PA} \vdash \mathrm{Con}_\tau \to \varphi$.

Let $d = d(T+\psi)$. Then, since ψ and Con_τ are Π_1, it follows from (3), (4), Lemma 3, and Theorem 6.4 that $a \cup d \geqslant b$. Suppose $a \leqslant d$. Then $T + \psi \vdash \mathrm{Con}_\tau$. But then, by (2) and (4), $T \vdash \varphi$, contradicting (1). Thus, $a \nleqslant d$. Let $c = d \cap b$. Then $c < b$. Finally, $a \cup c = (a \cup d) \cap (a \cup b) = b$. Thus, a is cupping. ∎

Theorem 14', below, is an improvement of Theorem 4.

A set G of degrees is *cofinal* in $\mathbf{D}_T$ if for every degree $a < 1_T$, there is a degree $b \in G$ such that $a \leqslant b < 1_T$.

Lemma 9. CON_T is cofinal in $\mathbf{D}_T$.

Proof. Suppose $b < 1_T$. By Corollary 2.5, even if T is not Σ_1-sound, there is a PR binumeration $\beta(x)$ of a theory of degree b such that $T + \mathrm{Con}_\beta$ is consistent. By Theorem 6.4, $b \leqslant d(T+\mathrm{Con}_\beta)$. By Theorem 2.8(b), there is a PR binumeration $\tau(x)$ of T such that $T \vdash \mathrm{Con}_\tau \leftrightarrow \mathrm{Con}_\beta$. Let $a = d(T+\mathrm{Con}_\tau)$. Then $b \leqslant a \in \mathrm{CON}_T$. ∎

Let P be a property of degrees. We shall say that there are *arbitrarily large* degrees having property P if the set of degrees having P is cofinal in $\mathbf{D}_T$. Every *sufficiently large* degree has P if for every degree $a < 1_T$, there is a b such that $a \leqslant b < 1_T$ and every degree c such that $b \leqslant c < 1_T$ has P.

If a is cupping and $a \leqslant b$, b is cupping. Thus, from Theorem 4 and Lemma 9 we get:

Corollary 3. *Every sufficiently large degree is a cupping degree.*

By Corollary 1, if T is Σ_1-sound, no degree, except 0_T and 1_T, has a complement whereas if T is not Σ_1-sound, some do. Also, of course, if $0_T \ll_\cap a < 1_T$, then a has no complement. But, even if a has no complement, it may still have a pseudocomplement (p.c.). For example, if $0_T \ll_\cap a$, 0_T is the p.c. of a. Clearly, if π is Π_1, then $d(T+\neg\pi)$ is the p.c. of $d(T+\pi)$. On the other hand we have the following:

Theorem 5. There is a degree which has no p.c.

The proof of this (and more) will be given in §3 (Theorem 17).

In addition to the usual (finitary) distributive laws, $\mathbf{D}_T$ also satisfies the following infinitary distributive laws. Let G be a set of degrees. $\cup G$ ($\cap G$) is then the l.u.b. (g.l.b.) of G, if it exists.

Theorem 6. (a) If $\cup G$ exists, then $a \cap \cup G = \cup\{a \cap b: b \in G\}$.
(b) If $\cap G$ exists, then $a \cup \cap G = \cap\{a \cup b: b \in G\}$.

By Theorem 6(a), if a has no p.c., then $\{b: a \cap b = 0_T\}$ has no l.u.b. In Lemma 25, below, we give a nontrivial example of a set G which has no g.l.b.

To prove Theorem 6(b) we need the following:

Lemma 10. The following conditions are equivalent:
(i) $A \uparrow B \geq C$.
(ii) For all (Σ_1) sentences χ and all m, if $A^T + \neg \mathrm{Con}_{C|m} \dashv_{\Sigma_1} T + \chi$, then $B \vdash \neg\chi$.

Proof. Suppose (i) holds. Let χ and m be such that $A^T + \neg\mathrm{Con}_{C|m} \dashv_{\Sigma_1} T + \chi$. There is a k such that $A^T + \mathrm{Con}_{B|k} \vdash \mathrm{Con}_{C|m}$. It follows that $T+\chi \vdash \neg\mathrm{Con}_{B|k}$, whence $B \vdash \neg\chi$. Thus, (ii) holds.

To prove that (ii) implies (i), suppose (i) fails, i.e., $A \uparrow B \not\geq C$. There is then an m such that for every k, $A^T + \mathrm{Con}_{B|k} \not\vdash \mathrm{Con}_{C|m}$. But then, by Theorem 4.3, there is a Σ_1 sentence χ such that $A^T + \neg\mathrm{Con}_{C|m} \dashv_{\Sigma_1} T+\chi$ and $T+\chi \not\vdash \neg\mathrm{Con}_{B|k}$ for every k. Since $\neg\chi$ is Π_1, it follows, by Lemma 2, that $B \not\vdash \neg\chi$. Thus, (ii) is false, as desired. ∎

Proof of Theorem 6. (a) Let $c = \cup G$. Clearly $a \cap c$ is an upper bound of $\{a \cap b: b \in G\}$. Suppose d is any upper bound of $\{a \cap b: b \in G\}$. It is then sufficient to show that $a \cap c \leq d$. Let $A \in a$, $C \in c$, $D \in d$. Then $A \downarrow B \leq D$ for every B such that $d(B) \in G$. But then, by Lemma 6, $B \leq D + \neg\mathrm{Con}_{A|k}$ for every such B and every k. It follows that for every k, $C \leq D + \neg \text{Con}_{A|k}$, whence, by Lemma 6, $A \downarrow C \leq D$ and so $a \cap c \leq d$. ♦

(b) Let $c = \cap G$. Clearly $a \cup c$ is a lower bound of $\{a \cup b: b \in G\}$. Suppose d is any lower bound of $\{a \cup b: b \in G\}$. It is then sufficient to show that $d \leq a \cup c$. Again let $A \in a$ etc. Then $D \leq A \uparrow B$ for every B such that $d(B) \in G$. But then, by Lemma 10, for every such B, every m and every Σ_1 sentence χ, if $A^T + \neg \text{Con}_{D|m} \dashv_{\Sigma_1} T + \chi$, then $B \vdash \neg \chi$. It follows that for every m and every Σ_1 sentence χ, if $A^T + \neg \text{Con}_{D|m} \dashv_{\Sigma_1} T + \chi$, then $C \vdash \neg \chi$. Hence, again by Lemma 10, $D \leq A \uparrow C$ and so $d \leq a \cup c$. ∎

Suppose $a \leq b$. Let [a,b] be the *interval* $\{c: a \leq c \leq b\}$. (We also write [a,b) for $\{c: a \leq c < b\}$ etc.) A natural (global) question concerning $\mathbf{D}_T$ is if all intervals [a,b], where $a < b < 1_T$, are isomorphic (in the obvious sense). The answer is negative.

If $c < d$, let $[d,c] = ([c,d], \geq)$. Another natural question is, under what conditions [a,b] is isomorphic to [d,c], where $a < b$ and $c < d$.

Theorem 7. (a) There are degrees a, $b \in (0_T, 1_T)$ such that the intervals $[0_T,a]$ and $[0_T,b]$ are not isomorphic.

(b) Suppose $a < b$ and $c < d$. Then [a,b] is not isomorphic to [d,c].

Theorem 7(a) follows at once from our next two lemmas.

The interval $[a_0,a_1]$, where $a_0 \leq a_1$, is said to satisfy the *reduction principle* (r.p.) if for any $b_0, b_1 \in [a_0,a_1]$, if $b_0 \cup b_1 = a_1$, there are $c_i \leq b_i$, $i = 0, 1$, such that $c_0 \cap c_1 = a_0$ and $c_0 \cup c_1 = a_1$. A degree a is *r.p.* if $[0_T,a]$ satisfies the r.p.

Lemma 11. If $a = d(T+\theta)$, where θ is Π_1, then a is r.p.

Proof. Let b_0, b_1 be such that $b_0 \cup b_1 = a$. There are then Π_1 sentences ψ_0, ψ_1 such that $d(T+\psi_i) \leq b_i$ and $T + \psi_0 \wedge \psi_1 \vdash \theta$. By Lemma 5.5, there are Π_1 sentences θ_0, θ_1 such that $T \vdash \theta_0 \vee \theta_1$, $T \vdash \psi_i \to \theta_i$, $i = 0, 1$, $T \vdash \theta_0 \wedge \theta_1 \to \psi_0 \wedge \psi_1$. Let $c_i = d(T+\theta_i)$, $i = 0, 1$. Then $c_i \leq b_i$, $c_0 \cap c_1 = 0_T$, and, by Lemma 3, $c_0 \cup c_1 = b_0 \cup b_1 = a$. ∎

Lemma 12. There is a degree $a < 1_T$ which is not r.p.

Proof. Let π be a Π_1 sentence undecidable in T. In case T is not Σ_1-sound we also need to assume that π is Σ_1-conservative over T (cf. Theorem 5.2). We now effectively define r.e. sets X_k of Π_1 sentences such that

(1) $T + X_k + \pi^i$ is consistent, $i = 0, 1$,
(2) $X_k \subseteq X_{k+1}$,
(3) $T + X_k + \pi \nvdash X_{k+1}$,
(4) $T + X_k + \neg\pi \leqslant T + X_{k+1}$.

Let $X_0 = \emptyset$. Then (1) holds for $k = 0$. Now suppose (1) holds for $k = n$. By (the proof of) Lemma 2.1, we can effectively find a Π_1 sentence ψ_n such that

(5) $T + X_n + \pi^i + \neg\psi_n^i$ is consistent, $i = 0, 1$.

Let $T_n =_{df} T + X_n + \neg\pi + \psi_n$. It follows that

(6) there is no Π_1 sentence θ such that $T_n \vdash \theta$ and $T + \theta \vdash \neg\pi$.

For suppose $T + \theta \vdash \neg\pi$. Then $T + \pi \vdash \neg\theta$ and so $T \vdash \neg\theta$, whence, by (5), $T_n \nvdash \theta$.

Let $X_{n+1} = Th(T_n) \cap \Pi_1$. Let $k = n+1$. Then (1) is satisfied for $i = 1$ and, by (6), (1) is satisfied for $i = 0$. Moreover (2) and (4) hold for $k = n$. Finally, $T + X_{n+1} \vdash \psi_n$ and so, by (5), (3) holds for $k = n$.

Let $a_0 = d(T + \cup\{X_k : k \in N\})$, $a_1 = d(T+\pi)$, and $a = a_0 \cup a_1$. Since $a_0 < 1_T$ and π is Σ_1-conservative over T, we have $a < 1_T$. We now show that a is not r.p. Let b_0 and b_1 be such that $b_0 \leqslant a_0$, $b_1 \leqslant a_1$, $b_0 \cap b_1 = 0_T$, and $b_0 \cup b_1 \geqslant a_1$. It is then sufficient to show that $b_0 \cup b_1 \not\geqslant a_0$.

Let $\theta_{i,k}$ be Π_1 sentences such that $b_i = d(T + \{\theta_{i,k} : k \in N\})$, $i = 0, 1$. We may assume that $T + \theta_{i,k+1} \vdash \theta_{i,k}$ for $i = 0, 1$ and all k. By Lemma 3, there is then an m such that $T + \theta_{0,m} \wedge \theta_{1,m} \vdash \pi$. $d(T+\theta_{0,m}) \leqslant b_0 \leqslant a_0$. Thus, by (2), there is an n such that $T + \theta_{0,m} \leqslant T + X_n$. Since $b_0 \cap b_1 = 0_T$, for every k, $T + \theta_{0,k} \vee \pi \leqslant T + \theta_{0,k} \vee (\theta_{0,m} \wedge \theta_{1,m}) \leqslant T + \theta_{0,m}$. It follows that $T + \theta_{0,k} \vee \pi \leqslant T + X_n$, whence $T + \theta_{0,k} \leqslant T + X_n + \neg\pi$ (cf. Corollary 6.3) and so, by (4), $T + \theta_{0,k} \leqslant T + X_{n+1}$. But this holds for all k, whence $b_0 \leqslant d(T+X_{n+1})$. Next, by (3), $b_0 \cup b_1 \leqslant b_0 \cup a_1 \leqslant d(T+X_{n+1}+\pi) \not\geqslant a_0$. It follows that $b_0 \cup b_1 \not\geqslant a_0$ and so the proof is complete. ∎

Proof of Theorem 7(b). Let $A \in a$ and let π be a Π_1 sentence such that $A \nvdash \pi$. Then $[a, d(A^T + \pi)]$ satisfies the r.p. (see the proof of Lemma 11). It follows that in $[a, b]$ there is a degree $e > a$ such that $[a, e]$ satisfies the r.p. Thus, it is sufficient to show that the dual of the r.p. is false in $[c, d]$ whenever $c < d$.

Let $C \in c$ and $D \in d$, and let π be such that $C \nvdash \pi$ and $D \vdash \pi$. Then, by Theorem 5.5(b) with $X = Th(C^T + \neg\pi)$, there are Σ_1 sentences σ_i, $i = 0, 1$, such that $C^T + \neg\sigma_i \equiv C^T + \sigma_{1-i}$, $i = 0, 1$, and $C^T + \neg\sigma_0 \wedge \neg\sigma_1 \nvdash \pi$. Let $c_i = d(C^T+\sigma_i) = d(C^T+\neg\sigma_{1-i})$, $i = 0, 1$. Then $c_0 \cap c_1 = c$ and $c_0 \cup c_1 \not\geqslant d$. Let $d_i = c_i \cap d$, $i = 0, 1$. Then $d_0 \cap d_1 = c$ and $d_0 \cup d_1 < d$. Suppose now $d_i \leqslant e_i \leqslant d$, $i = 0, 1$, and $e_0 \cap e_1 = c$. We have to show that $e_0 \cup e_1 < d$. Let $E_0 \in e_0$. $c_1 \cap e_0$

$= c_1 \cap d \cap e_0 = d_1 \cap e_0 \leq e_1 \cap e_0 = c$. It follows that $(C^T + \neg \sigma_0) \downarrow E_0 \leq C^T$. But then, by Lemma 6, for every Π_1 sentence θ, if $E_0 \vdash \theta$, then $C^T + \neg \sigma_0 \leq C^T + \neg \theta$, whence $C^T + \sigma_0 \vdash \theta$. It follows that $e_0 \leq c_0$ and so $e_0 = d_0$. Similarly, $e_1 = d_1$. Hence $e_0 \cup e_1 = d_0 \cup d_1 < d$ and the proof is complete. ∎

Theorem 7(a) leads to the problem of determining the exact number of nonisomorphic intervals of D_T. This problem remains open.

We have actually proved more than is stated in Theorem 7. Let $L = \{\leq, \cap, \cup, 0, 1\}$ be the language of the theory of lattices with a bottom and a top element. Formulated in L, the r.p. is an $\forall \exists$ sentence. Hence, by the proof of Theorem 7(a), there are degrees a, $b \in (0_T, 1_T)$ and an $\forall \exists$ sentence of L which holds in $[0_T, a]$ but not in $[0_T, b]$. (This is, so far, the only known way of proving that two intervals of $\mathbf{D}_T$ are nontrivially nonisomorphic.) Similarly, the proof of Theorem 7(b) shows that if $a < b$ and $c < d$, there is an $\exists \forall \exists$ sentence which is true in [a,b] and false in [d,c].

§2. A classification of degrees. When there is no risk of confusion we shall use φ and X in place of $T + \varphi$ and $T + X$. Thus, $d(\varphi)$ is $d(T+\varphi)$, $X < \varphi$ means that $T + X < T + \varphi$, $\varphi \equiv \psi$ that $T + \varphi \equiv T + \psi$, etc. We also write $a \ll b$ to mean that $a \ll_\cap b$. Note that if $a \ll b$ and $a \ll c$, then $a \ll b \cap c$. $A \ll B$ means that $d(A) \ll d(B)$. In what follows σ, σ_0, etc. will be used to denote Σ_1 sentences and π, π_0, etc. to denote Π_1 sentences.

A degree a is Φ if there is a Φ sentence φ such that $a = d(\varphi)$. By the proof of Theorem 6.11(a), it is clear that every degree is Π_2 and Σ_2. This can be somewhat improved:

Theorem 8. Every degree is Δ_2.

Proof. Let a be any degree. There is a primitive recursive set X of Π_1 sentences such that $a = d(X)$. Let $\xi(x)$ be a PR binumeration of X and let φ be such that
$$PA \vdash \varphi \leftrightarrow \forall z([\Pi_1](\varphi, z) \to (\xi(z) \to Tr_{\Pi_1}(z))).$$
Then φ is Π_2 and $T + \varphi$ is a Π_1-conservative extension of $T + X$ (cf. the proof of Theorem 5.4(a)). It follows that $a = d(\varphi)$. Using Lemma 5.1(i) and Lemma 1.3(v) (applied to $\neg \varphi$), we get:
$$PA \vdash \varphi \leftrightarrow \forall z(\xi(z) \to Tr_{\Pi_1}(z)) \vee$$
$$\exists z(\neg[\Pi_1](\varphi, z) \wedge \forall u < z(\xi(u) \to Tr_{\Pi_1}(u))).$$
Thus, φ is Δ_2. ∎

By Theorem 8, in terms of the arithmetical hierarchy, the only interesting (proper) subsets of D_T are the sets of B_1 degrees, Σ_1 degrees, Π_1 degrees, and degrees which are both Σ_1 and Π_1. (If T is not Σ_1-sound, there are also Δ_1^T degrees other than 0_T and 1_T; see e.g. the proof of Corollary 1.) The object of the rest of this § is simply to show that these sets are different and that there is a non-B_1 degree. More detailed information about the Σ_1 and the Π_1 degrees will be given in the next §.

Our next lemma is a restatement of Theorem 6.11(b).

Lemma 13. If X is an r.e. set of Σ_1 sentences, then $d(X)$ is Σ_1.

It is occasionally useful to know that there are sentences σ_k, σ satisfying the following condition:
(#) $\langle \sigma_k \rangle_{k<\omega}$ is (primitive) recursive, for all k, $T + \sigma_{k+1} \vdash \sigma_k$ and
$\sigma_k < \sigma_{k+1}$, and $\sigma \equiv \{\sigma_k : k \in N\}$.
This follows at once from (the proof of) Lemma 2.1 (applied to the sets $\{\varphi : Q + \varphi \leqslant T + \sigma_k\}$; $\sigma_0 := 0 = 0$) and Lemma 13.

Theorem 9. (a) There is a Π_1 degree which isn't Σ_1.
  (b) There is a Σ_1 degree which isn't Π_1.
  (c) There is a degree other than 0_T and 1_T which is both Σ_1 and Π_1.
  (d) There is a B_1 degree which is neither Σ_1 nor Π_1.
  (e) There is a degree which isn't B_1.

Proof. (a) Let π be such that $\neg\pi$ is Π_1-conservative over T and $T \nvdash \neg\pi$. Then, by Lemma 8, $0_T \ll d(\pi) < 1_T$ and so, by Lemma 7, $d(\pi)$ is not Σ_1. ♦

(b) Let $\langle \sigma_k \rangle_{k<\omega}$ and σ be as in (#). Suppose $d(\sigma)$ is Π_1 and let π be such that $\sigma \equiv \pi$. Then $\pi \equiv \{\sigma_k : k \in N\}$ and so there is an m such that $T + \sigma_m \vdash \pi$. But then $\{\sigma_k : k \in N\} \leqslant \sigma_m$, contradicting (#). Thus, $d(\sigma)$ isn't Π_1. ♦

(d) The easiest way to prove this is to define π as in the proof of (a) and then σ as in the proof of (b), except that T is replaced by $T + \pi$. Then $d(\pi \wedge \sigma)$ is neither Σ_1 nor Π_1. Details are left to the reader. ♦

Theorem 9(b) can be strengthened in several different ways; see, for example, Theorems 14, 16, 17.

Theorem 9(c) will be derived from the following lemma, which will also be used later.

Lemma 14. There are Π_1 sentences θ_i, $i = 0, 1$, such that
(i) $T \nvdash \theta_i$,
(ii) $PA \vdash \theta_0 \vee \theta_1$,

(iii) $PA \vdash \theta_0 \wedge \theta_1 \rightarrow Con_T$,
(iv) $PA + Con_T \vdash \neg Pr_T(\theta_i)$,
(v) $PA + Con_T \vdash \theta_i$,
(vi) $\theta_i \equiv \neg \theta_{1-i}$.

Proof. Let θ_i, $i = 0, 1$, be such that
$$PA \vdash \theta_i \leftrightarrow \forall z(Prf_T(\theta_i, z) \rightarrow \exists u < z + i Prf_T(\theta_{1-i}, u)).$$
A standard argument proves (i). Formalizing this argument in PA we get (iv). (ii) and (iii) are immediate. (v) follows from (iv). By (ii),
(1) $\quad PA \vdash Pr_T(\neg \theta_i) \rightarrow Pr_T(\theta_{1-i})$.
Also,
(2) $\quad PA \vdash \neg \theta_{1-i} \leftrightarrow Pr_T(\theta_{1-i}) \wedge \theta_i$.
By Theorem 6.8, $\theta_i \wedge Pr_T(\neg \theta_i) \leqslant \theta_i$. By (1), it follows that $\theta_i \wedge Pr_T(\theta_{1-i}) \leqslant \theta_i$ and so, by (2), $\neg \theta_{1-i} \leqslant \theta_i$. But then, by (ii), $\theta_i \equiv \neg \theta_{1-i}$, i.e., (vi) holds. ∎

Proof of Theorem 9(c). Let θ_i be as in Lemma 14. Let $a = d(\theta_0)$. Then a is Π_1 and, by Lemma 14(vi), a is Σ_1. By Lemma 14(i), $a > 0_T$. Finally, by Lemma 14(i) and (ii), $a < 1_T$. ♦

To prove Theorem 9(e) we need the following:

Lemma 15. Suppose X is r.e. and for every k, $X|k \ll X$. Then if φ is B_1 and $X \leqslant \varphi$, then $X \ll \varphi$. Thus, *a fortiori* $d(X)$ is not B_1.

Proof. φ can be written in the form $(\pi_0 \wedge \sigma_0) \vee ... \vee (\pi_n \wedge \sigma_n)$. It is then sufficient to show that if $X \leqslant \pi \wedge \sigma$, then $X \ll \pi \wedge \sigma$. Let χ be a Σ_1 sentence such that $\pi \wedge \sigma \leqslant X + \chi$. Then, by Lemma 7, it suffices to show that $T + X \vdash \neg \chi$. By assumption, there is a k such that $T + X|k + \chi \vdash \pi$. Hence $T + \pi \wedge \sigma \dashv T + X|k + (\chi \wedge \sigma)$ and so $X \leqslant X|k + (\chi \wedge \sigma)$. But then, since $X|k \ll X$, by Lemma 7, $T + X \vdash \neg(\chi \wedge \sigma)$. But $X \leqslant \pi \wedge \sigma$. It follows that $T + \pi \wedge \sigma \vdash \neg \chi$, whence $T + X + \chi \vdash \neg \chi$ and so $T + X \vdash \neg \chi$, as was to be shown. ∎

Proof of Theorem 9(e). By (the proof of) Theorem 5.2, we can effectively construct sentences π_n such that $\neg \pi_n$ is Π_1-conservative over but not provable in $T + \{\pi_k : k < n\}$. Let $X = \{\pi_k : k \in N\}$. Then, by Lemma 8, $X|k \ll X$ for all k. So, by Lemma 15, $d(X)$ is not B_1. ∎

Theorem 13, below, is an improvement of Theorem 9(e).

§3. Σ_1 and Π_1 degrees.

This § is devoted to a discussion of the Σ_1 and Π_1 degrees and the relations between them.

The l.u.b. of two Π_1 degrees is Π_1 and the g.l.b. of two Π_1 (Σ_1) degrees is Π_1 (Σ_1).

Let us say that a is *high* if a $\gg 0_T$, *low* otherwise. Thus, by Lemma 7, a is low iff there is a Σ_1 degree b such that a $\leq$ b $< 1_T$. By Lemma 8, if $\neg\pi$ is Π_1-conservative over T, $d(\pi)$ is high. By Corollary 2, every member of CON_T is high.

The following lemma is sometimes useful.

Lemma 16. Suppose a is high. Then for any b, [a $\cap$ b,b) contains no Σ_1 degree; in fact, if c is Σ_1 and a $\cap$ b $\leq$ c, then b $\leq$ c.

Proof. Let $A \in a$, $B \in b$, and $c = d(\sigma)$. Suppose $A \downarrow B \leq T + \sigma$. Then, by Lemma 6, $A \leq T + \sigma \wedge \neg\mathrm{Con}_{B|k}$ for every k. Since a is high, it follows, by Lemma 7, that $T + \sigma \vdash \mathrm{Con}_{B|k}$, for every k, and so $B \leq T + \sigma$. ∎

Theorem 10. (a) The set of Π_1 degrees is cofinal in $\mathbf{D}_T$.

(b) The set of Σ_1 degrees is not cofinal in $\mathbf{D}_T$; in fact, for every degree a $> 0_T$, there is a degree b $<$ a such that [b,a) contains no Σ_1 degree.

(c) There is a low Π_1 degree which is not Σ_1.

Proof. (a) Since all members of CON_T are Π_1, this follows from Lemma 9. ♦

(b) By Theorem 3(b), there is a high degree c such a $\not\leq$ c. Let b = c $\cap$ a. Then, by Lemma 16, b is as desired. ♦

(c) Let a be any low Π_1 degree $> 0_T$. By Theorem 3(b), there is a high Π_1 degree c $\not\geq$ a. Let b = a $\cap$ c. Then b is low and and Π_1. Finally, by Lemma 16, b is not Σ_1. ∎

Using Theorem 2 we can now prove the following corollary.

Corollary 4. (a) Suppose a is not Π_1 and a $\in$ (b,c). There are then degrees b', c' such that a $\in$ (b',c') $\subseteq$ (b,c) and [b',c'] contains no Π_1 degree.

(b) Suppose a is not Σ_1 and a $\in$ (b,c). There are then degrees b', c' such that a $\in$ (b',c') $\subseteq$ (b,c) and [b',c'] contains no Σ_1 degree.

Proof. (a) By Theorem 2, there are degrees b_0, b_1 such that b $\leq b_i <$ a, i = 0, 1, and $b_0 \cup b_1$ = a. Either [b_0,a] or [b_1,a] contains no Π_1 degree. If not, a would be the l.u.b. of two Π_1 degrees and therefore Π_1. Suppose [b_i,a] contains no Π_1 degree and let b' = b_i. By Theorem 2, there are degrees c_0, c_1 such that a $< c_i \leq$ c, i = 0, 1, and $c_0 \cap c_1$ = a. Either [b',c_0] or [b',c_1] contains no Π_1 degree. For suppose $d_i \in$ [b',c_i] and d_i is Π_1, i = 0, 1. Then

$d_0 \cap d_1 \in [b',a]$ and $d_0 \cap d_1$ is Π_1, a contradiction. Suppose $[b',c_j]$ contains no Π_1 degree and let $c' = c_j$. Then b' and c' are as desired. ◆

(b) By a slight modification of the proof of Theorem 10(b), which we leave to the reader, there is a degree b' such that $b \leq b' < a$ and $[b',a]$ contains no Σ_1 degree. The rest of the proof is the same as in the proof of (a). ■

Theorem 10(b) leads to the question if there are arbitrarily small Σ_1 degrees. By our next result, the answer is affirmative; later we shall prove a stronger result (Theorem 15).

Theorem 11. If $0_T < a$, there is a Σ_1 and Π_1 degree $b \in (0_T, a)$.

To prove this we need a lemma on partial conservativity.

Lemma 17. Let X be an r.e. set. There is then a PR formula $\eta(y,x,z)$ such that for all k and θ,
(i) if $k \in X$, then $T + \theta \vdash \neg \exists z \eta(\theta,k,z)$,
(ii) if $k \notin X$, then $\exists z \eta(\theta,k,z)$ is Π_1-conservative over $T + \theta$.

The proof of Lemma 17 is similar to the proof of Lemma 5.3 (for $\Gamma = \Sigma_1$) and is left to the reader.

Proof of Theorem 11. Let $\forall u \delta(u)$, where $\delta(u)$ is PR, be a Π_1 sentence such that $0_T < d(\forall u \delta(u)) < a$. By Lemma 17, there is a PR formula $\gamma(x,z)$ such that
(1)   if $T \vdash \varphi$, then $T \vdash \neg \exists z \gamma(\varphi,z)$,
(2)   if $T \nvdash \varphi$, then $\exists z \gamma(\varphi,z)$ is Π_1-conservative over $T + \varphi$.
Let θ be such that
(3) $PA \vdash \theta \leftrightarrow \forall u(\neg \delta(u) \to \exists z < u \gamma(\theta,z))$,
and let
$$\sigma := \exists z(\gamma(\theta,z) \wedge \forall u \leq z \delta(u)).$$
Then
(4) $PA \vdash \sigma \leftrightarrow \exists z \gamma(\theta,z) \wedge \theta$,
(5) $PA + \theta + \neg \exists z \gamma(\theta,z) \vdash \forall u \delta(u)$.
It follows that
(6) $T \nvdash \theta$.
For suppose not. Then, by (1), $T \vdash \neg \exists z \gamma(\theta,z)$ and so, by (5), $T \vdash \forall u \delta(u)$, contrary to the choice of $\delta(u)$.

By (3), $\theta \leq \forall u \delta(u)$ and so $d(\theta) < a$. By (6), $0_T < d(\theta)$. Finally, by (4), (6), (2), $\sigma \equiv \theta$. Thus, $b = d(\sigma)$ is as claimed. ■

It is natural to ask if D_T is "generated" by some "small" set of degrees, for example, the set of Σ_1 degrees. We prove two negative results, Theorems 12 and 13, and one partial positive result, Theorem 14 (and 14').

Let $\text{Cl}^{\cup}(\Phi)$ ($\text{Cl}(\Phi)$) be the set of degrees obtained from the Φ degrees by closing under $\cup$ ($\cup$ and $\cap$). Obviously, the set of Π_1 degrees is closed under $\cup$ and $\cap$. If Φ is Σ_1 or B_1, the set of Φ degrees is closed under $\cap$ and so $\text{Cl}(\Phi) = \text{Cl}^{\cup}(\Phi)$. By Lemma 14, $\text{CON}_T \subseteq \text{Cl}(\Sigma_1)$.

Theorem 12. There is a Π_1 degree not in $\text{Cl}(\Sigma_1)$.

This is an immediate consequence of the following two lemmas.

Lemma 18. If $a \in \text{Cl}(\Sigma_1)$, there is a smallest Σ_1 degree $\geqslant a$.

Proof. Suppose $a \in \text{Cl}(\Sigma_1)$. There are then Σ_1 degrees $d(\sigma_i)$, $i \leqslant n$, such that $a = d(\sigma_0) \cup \ldots \cup d(\sigma_n)$. But then $d(\sigma_0 \wedge \ldots \wedge \sigma_n)$ is the smallest Σ_1 degree $\geqslant a$. This can be seen as follows. Suppose $d(\sigma_0) \cup \ldots \cup d(\sigma_n) \leqslant d(\sigma)$. Let π be such that $T + \sigma_0 \wedge \ldots \wedge \sigma_n \vdash \pi$. Then $T + \sigma_0 \vdash \sigma_1 \wedge \ldots \wedge \sigma_n \to \pi$. Now, $\sigma_1 \wedge \ldots \wedge \sigma_n \to \pi$ is a Π_1 sentence. It follows that $T + \sigma \vdash \sigma_1 \wedge \ldots \wedge \sigma_n \to \pi$. But then $T + \sigma_1 \vdash \sigma \wedge \sigma_2 \wedge \ldots \wedge \sigma_n \to \pi$ and so $T + \sigma \vdash \sigma_2 \wedge \ldots \wedge \sigma_n \to \pi$. Continuing in this way we eventually get $T + \sigma \vdash \pi$, as desired. ∎

Lemma 19. There is a Π_1 degree a for which there is no smallest Σ_1 degree $\geqslant a$.

Proof. Let $\langle \sigma_k \rangle_{k<\omega}$ and σ be as in (#). Let $a = d(\neg\sigma)$. Then a is Π_1. Now let χ be any Σ_1 sentence such that $a \leqslant d(\chi)$. Then $T + \chi \vdash \neg\sigma$ and so $T + \sigma \vdash \neg\chi$. It follows that there is a k such that $T + \sigma_k \vdash \neg\chi$ and so
(1) $T + \chi \vdash \neg\sigma_k$.
Since $\sigma_k < \sigma$, there is a sentence π such that $T + \sigma \vdash \pi$ and $T + \sigma_k \nvdash \pi$. It follows that
(2) $T + \neg\pi \vdash \neg\sigma$,
(3) $T + \neg\pi \nvdash \neg\sigma_k$.
But then, by (2), $a \leqslant d(\neg\pi)$ and, by (1) and (3), $\chi \nleqslant \neg\pi$. Thus, $d(\chi)$ is not the smallest Σ_1 degree $\geqslant a$. ∎

A strengthening of Lemma 19 will be proved later (Lemma 25).

Theorem 13. $\text{Cl}(B_1) \neq D_T$.

This follows at once from Lemmas 20 and 23, below. We write $a \lll b$ to

§3. Σ_1 and Π_1 Degrees   133

mean that if $A \in a$, $B \in b$, then $A < B$ and for every set X of Σ_1 sentences (not necessarily r.e.), if $B \dashv_{\Pi_1} A + X$, then $A + X$ is inconsistent. (If π is Π_1 and $\neg\pi$ is Π_1-conservative over A, then $A <<< A + \pi$. By Lemma 7, if $a <<< b$, then $a << b$.) Let $\Sigma_1 \wedge \Pi_1$ be the set of sentences of the form $\sigma \wedge \pi$.

Lemma 20. If $a \in Cl(B_1)$ and $0_T << a$, then $0_T <<< a$.

In the proof of Lemma 20 we use the following observation.

Lemma 21. If $a <<< b$, c, then $a <<< b \cap c$.

Proof. Let $A \in a$, $B \in b$, $C \in c$. We use π and θ to denote Π_1 sentences provable in B and C, respectively, Let X be any set of Σ_1 sentences and suppose $B \downarrow C \dashv_{\Pi_1} A + X$. Then $A + X \vdash \pi \vee \theta$ and so $A + X + \neg\pi \vdash \theta$. Since $a <<< c$, it follows that $A + X \vdash \pi$ and so, since $a <<< b$, $A + X$ is inconsistent. ∎

Proof of Lemma 20. Suppose $a \in Cl(B_1)$ and $0_T << a$. There are degrees $a_i \in Cl^{\cup}(\Sigma_1 \wedge \Pi_1)$ such that $a = a_0 \cap ... \cap a_k$. Clearly, $a_i >> 0_T$, $i \leq k$. Thus, by Lemma 21, it is sufficient to show that
(1)   if $b \in Cl^{\cup}(\Sigma_1 \wedge \Pi_1)$ and $0_T << b$, then $0_T <<< b$.
Let $\sigma_0,...,\sigma_n$ and $\pi_0,...,\pi_n$ be such that $b = d(\sigma_0 \wedge \pi_0) \cup ... \cup d(\sigma_n \wedge \pi_n)$. Let $B \in b$. Then $B \vdash \pi_0 \wedge ... \wedge \pi_n$ and
(2) $T + \sigma_i \leq B$ for $i \leq n$.
Let X be any set of Σ_1 sentences such that
(3) $B \dashv_{\Pi_1} T + X$.
Then
(4) $T + X \vdash \neg\sigma_0 \vee ... \vee \neg\sigma_n$.
For, by (3), there is a k such that $T + \wedge X | k \vdash \pi_0 \wedge ... \wedge \pi_n$. It follows that $b \leq c$, where $c = d(\wedge X | k \wedge \sigma_0 \wedge ... \wedge \sigma_n)$. Since $b >> 0_T$ and c is Σ_1, by Lemma 7, this implies that $c = 1_T$, whence (4) follows.

By (4), there is a k_0 such that $T + \sigma_0 \vdash \neg \wedge X | k_0 \vee \neg\sigma_1 \vee ... \vee \neg\sigma_n$. But then, by (2) and (3), $T + X \vdash \neg\sigma_1 \vee ... \vee \neg\sigma_n$. Continuing in this way we eventually obtain the conclusion that $T + X$ is inconsistent.

This proves (1) and so concludes the proof of Lemma 20. ∎

Lemma 22. For every Σ_3^0 set X, there is a formula $\xi(x)$ such that
    $X = \{k: T <<< T + \xi(k)\}$.

Proof. Let $S(k,p,q)$ be an r.e. relation such that $X = \{k: \exists p \forall q S(k,p,q)\}$. By Lemma 3.2(b), there are a Π_1 formula $\sigma(x,z,u)$ and a Σ_1 formula $\sigma'(x,z,u)$

such that
(1) $T \vdash \sigma'(x,z,u) \to \sigma(x,z,u)$,
(2) if $S(k,p,q)$, then $T \vdash \sigma'(k,p,q)$,
(3) $T + Y$ is consistent where $Y = \{\neg\sigma(k,p,q): \text{not } S(k,p,q)\}$.

By (2) and Lemma 5.2, there is a Σ_1 formula $\rho(x,y)$ such that
(4) $T \vdash \rho(k,p) \to \sigma'(k,p,q)$,
(5) if $\forall q S(k,p,q)$, then $\rho(k,p)$ is Π_1-conservative over T.

By Theorem 5.4(b), there is a formula $\xi(x)$ such that
(6) $T + \xi(k)$ is a Π_1-conservative extension of $T + \{\neg\rho(k,p): p \in N\}$.

Now, suppose $k \in X$. Let p be such that $\forall q S(k,p,q)$. By (5), $\rho(k,p)$ is Π_1-conservative over T. By (6), $T + \xi(k) \vdash \neg\rho(k,p)$. It follows that $T <<< T + \xi(k)$.

Next, suppose $k \notin X$. Let Y be as in (3). To show that not $T <<< T + \xi(k)$ it is sufficient to show that
$$T + \xi(k) \dashv_{\Pi_1} T + Y.$$
Suppose $T + \xi(k) \vdash \pi$. Then, by (6),
(7) $T + \{\neg\rho(k,p): p \in N\} \vdash \pi$.
Since $k \notin X$, for each p there is a q_p such that not $S(k,p,q_p)$. By (4) and (7), $T + \{\neg\sigma'(k,p,q_p): p \in N\} \vdash \pi$. And so, by (1), $T + Y \vdash \pi$, as desired. ∎

Lemma 23. There is a degree a such that $a \gg 0_T$ and not $a \ggg 0_T$.

Proof. Let $Y = \{\varphi: T << T + \varphi\}$ and $Z = \{\varphi: T <<< T + \varphi\}$. Then $Z \subseteq Y$. By the defintion of $<<$, Y is Π_3^0. By Lemma 22, Z is not Π_3^0. Thus, $Y \neq Z$ and so $Y \nsubseteq Z$. Let $\theta \in Y - Z$. (For example, let $\xi(x)$ be as in Lemma 22 with $X = Y^c$ and let θ be such that $PA \vdash \theta \leftrightarrow \xi(\theta)$.) Then $a = d(\theta)$ is as desired. ∎

The degree mentioned in Lemma 19 cannot be arbitrarily large: if a is high, there is a smallest Σ_1 degree $\geq a$, namely 1_T. Similarly, the degree a defined in the proof of Theorem 13 (Lemma 23) cannot be arbitrarily large; it is not $\ggg 0_T$. This is explained, at least partially, by the following surprising:

Theorem 14. (a) Every sufficiently large degree is the l.u.b. of a Σ_1 degree and a Π_1 degree.
(b) Every sufficiently large degree is the l.u.b. of two Σ_1 degrees.

Proof. We may assume that $d(\text{Con}_T) < 1_T$. By Lemma 9, it is sufficient to consider degrees a such that $d(\text{Con}_T) \leq a < 1_T$. Let $\pi_n := \forall u \delta_n(u)$, where $\delta_n(u)$ is PR, be Π_1 sentences such that $a = d(\{\pi_n : n \in N\})$. We may assume

§3. Σ_1 and Π_1 Degrees

that for all n,
(1) $T \vdash \pi_0 \to \text{Con}_T$,
(2) $T \vdash \pi_{n+1} \to \pi_n$.
   (a) We define Π_1 sentences φ_n and ψ_n in the following way:
(3) $T \vdash \varphi_n \leftrightarrow \forall z(\text{Prf}_T(\bigvee\{\varphi_k:k\leq n\},z) \to \exists u\leq z \neg \delta_{n+1}(u))$,
   $\psi_n := \forall u(\neg \delta_{n+1}(u) \to \exists z<u \text{Prf}_T(\bigvee\{\varphi_k:k\leq n\},z))$.
It follows that
(4) $T \vdash \varphi_n \vee \psi_n$,
(5) $T + \varphi_n \wedge \psi_n \vdash \neg\text{Pr}_T(\bigvee\{\varphi_k:k\leq n\}) \wedge \pi_{n+1}$,
(6) $T + \pi_{n+1} \vdash \psi_n$,
(7) $T + \neg\varphi_n \vdash \text{Pr}_T(\bigvee\{\varphi_k:k<n\})$,
(8) $\neg\bigvee\{\varphi_k: k<n\} \leq \pi_n$.
($\bigvee\{\varphi_k: k<0\} := \bot$.) (4), (5), (6) are standard.
   Since $\neg\varphi_n$ is Σ_1, we have $T + \neg\varphi_n \vdash \text{Pr}_T(\neg\varphi_n)$. Also, by (3),
   $T + \neg\varphi_n \vdash \text{Pr}_T(\bigvee\{\varphi_k:k\leq n\})$.
But then (7) follows.
   By Theorem 6.4, (8) follows from
(9) $T + \pi_n \vdash \neg\text{Pr}_T(\bigvee\{\varphi_k:k<n\})$.
By (1), (9) holds for n = 0. Suppose (9) holds for n = m. To show that it holds for n = m+1, we argue in T as follows: "Suppose π_{m+1}. Then, by (6), ψ_m. Also, by (2) and the inductive assumption, $\neg\text{Pr}_T(\bigvee\{\varphi_k:k<m\})$ and so, by (7), φ_m. Finally, by (5), $\neg\text{Pr}_T(\bigvee\{\varphi_k:k<m+1\})$, as desired." Thus, (9) holds for n = m+1. This proves (9) and so we have proved (8).
   Next we show that for all n,
(10) $T + \bigwedge\{\psi_k: k<n\} + \text{Con}_T \vdash \varphi_n$.
We first show that
(11) $T + \text{Con}_T \vdash \varphi_0$,
(12) $T + \psi_n + \varphi_n \vdash \varphi_{n+1}$.
(11) follows from (7) with n = 0. (12) follows from (5) and (7).
   Now (10) follows from (11) and (12).
   Let $a_0 = d(\{\neg\varphi_k:k\in \mathbb{N}\})$, $a_1 = d(\text{Con}_T)$, $a_2 = d(\{\varphi_k:k\in \mathbb{N}\})$. Then, by Lemma 13, a_0 is Σ_1. a_1 is Π_1. By (8) and Orey's compactness theorem, $a_0 \leq a$ and, by hypothesis, $a_1 \leq a$. But then $a_0 \cup a_1 \leq a$. By (4) and (5), $a_0 \cup a_2 \geq a$. By (4) and (10), $a_0 \cup a_1 \geq a_2$. It follows that $a_0 \cup a_1 \geq a$ and so $a_0 \cup a_1 = a$, as desired. ♦
   (b) Let θ_i, i = 0, 1, be as in Lemma 14. We define Π_1 sentences φ_n and ψ_n in the following way:
(13) $T \vdash \varphi_n \leftrightarrow \forall z(\text{Prf}_T(\theta_0 \vee \bigvee\{\varphi_k:k\leq n\},z) \to \exists u\leq z\neg\delta_{n+1}(u))$,
   $\psi_n := \forall u(\neg\delta_{n+1}(u) \to \exists z<u \text{Prf}_T(\theta_0 \vee \bigvee\{\varphi_k:k\leq n\},z))$.

It follows that
(14) $T \vdash \varphi_n \vee \psi_n$,
(15) $T + \varphi_n \wedge \psi_n \vdash \neg \Pr_T(\theta_0 \vee \bigvee\{\varphi_k : k \leq n\}) \wedge \pi_{n+1}$,
(16) $T + \pi_{n+1} \vdash \psi_n$,
(17) $T + \neg \varphi_n \vdash \Pr_T(\theta_0 \vee \bigvee\{\varphi_k : k < n\})$,
(18) $\neg \theta_0 \wedge \neg \bigvee\{\varphi_k : k < n\} \leq \pi_n$.

The proofs of (14)–(18) are almost the same as the proofs of (4)–(8).

Next we show that for all n,
(19) $T + \bigwedge\{\psi_k : k < n\} + \theta_0 \wedge \theta_1 \vdash \varphi_n$.

We first show that
(20) $T + \theta_0 \wedge \theta_1 \vdash \varphi_0$,
(21) $T + \psi_n + \varphi_n \vdash \varphi_{n+1}$.

(20) follows from Lemma 14(iii) and (iv) and (17) with n = 0. (21) follows from (15) and (17).

Now (19) follows from (20) and (21).

Let $a_0 = d(\neg \theta_0 + \{\neg \varphi_k : k \in N\})$, $a_1 = d(\theta_0)$. Then a_0 is Σ_1. By Lemma 14(vi), a_1 is Σ_1. By (18), $a_0 \leq a$ and, by Lemma 14(v), $a_1 \leq a$. But then $a_0 \cup a_1 \leq a$. By Lemma 14(ii), (14), (19), and (15), $a_0 \cup a_1 \geq a$. Thus, $a_0 \cup a_1 = a$, as desired. ∎

The proof of Theorem 14 actually yields the following stronger result; Theorem 14'(a) is also an improvement of Theorem 4.

Theorem 14'. (a) Suppose $a \in \text{CON}_T$ and $a \leq b < 1_T$. There is then a Σ_1 degree c such that $a \cup c = b$.

(b) Suppose $a \in \text{CON}_T$. There are then degrees a_0, a_1 such that (i) a_0 and a_1 are both Σ_1 and Π_1, (ii) $a_0 \cap a_1 = 0_T$, (iii) $a_0 \cup a_1 = a$, (iv) for every degree $b \geq a$, there is a Σ_1 degree b_i such that $a_i \cup b_i = b$, i = 0, 1.

One way to strengthen Theorem 12 would be to show that there is a Π_1 degree $a > 0_T$ such that no Σ_1 degree cups to a. This, however, is not the case:

Theorem 15. For every Π_1 degree $a > 0_T$, there is a Σ_1 (and Π_1) degree which cups to a.

Proof. The following proof is similar to that of Theorem 11. Let π be such that $a = d(\pi)$ and let $\delta(u)$ be a PR formula such that $\pi := \forall u \delta(u)$. By Lemma 17, there is a PR formula $\eta(x,y,z)$ such that for all φ, ψ,
(1) if $T + \varphi \vdash \pi$, then $T + \psi \vdash \neg \exists z \eta(\varphi, \psi, z)$,

§3. Σ_1 and Π_1 Degrees

(2) if $T + \varphi \nvdash \pi$, then $\exists z \eta(\varphi,\psi,z)$ is Π_1-conservative over $T + \psi$.

Next let θ and χ be such that

(3) $T \vdash \theta \leftrightarrow \forall u(\neg\delta(u) \rightarrow \exists z < u \eta(\chi,\theta,z))$,

$\quad\quad T \vdash \chi \leftrightarrow \forall z(\eta(\chi,\theta,z) \rightarrow \exists u \leqslant z \neg\delta(u))$.

Then

(4) $T \vdash \theta \vee \chi$,

(5) $T \vdash (\theta \wedge \chi) \rightarrow \pi$.

We now show that

(6) $T + \chi \nvdash \pi$.

Suppose not. Then, by (1) and (3), $T + \theta \vdash \pi$. But then, by (4), $T \vdash \pi$, contrary to assumption. This proves (6).

Now let

$\quad\quad \sigma := \exists z(\eta(\chi,\theta,z) \wedge \forall u \leqslant z \delta(u))$.

Then

$\quad\quad T \vdash \sigma \leftrightarrow \exists z \eta(\chi,\theta,z) \wedge \theta$.

By (3), $d(\theta) \leqslant a$. By (2) and (6), $\sigma \equiv \theta$. Thus, $d(\sigma)$ is Σ_1 and Π_1. Let $b = a \cap d(\chi)$. By (6), $a \nleqslant d(\chi)$ and so $b < a$. By (5), $d(\sigma) \cup b = a$. Thus, $d(\sigma)$ cups to a. Also note that b (is Π_1 and) $d(\sigma) \cap b = 0_T$. ∎

The problem if for every degree $a > 0_T$, there is a Σ_1 degree which cups to a remains open. By Theorem 14, this is true of every sufficiently large degree.

Our next task is to show that the result of interchanging Σ_1 and Π_1 in Theorem 15 is false.

Theorem 16. There is a Σ_1 degree $a > 0_T$ such that no Π_1 degree cups to a.

Let $\xi(x)$ be as in Lemma 5.8 with $n = 1$ and let $a = d(\{\xi(k) : k \in N\})$. Then $a > 0_T$ and no Π_1 degree cups to a (see the proof of Theorem 3(a)). To obtain a Σ_1 degree satisfying these conditions we first prove the following refinement of Lemma 5.8 (for $n = 1$).

Lemma 24. There are Π_1 formulas $\xi(x)$, $\eta(x)$ and Σ_1 sentences χ_k such that
(i) $T \nvdash \xi(k)$,
(ii) $T \vdash \eta(k) \rightarrow \xi(k)$,
(iii) $T \vdash \xi(k+1) \rightarrow \eta(k)$,
(iv) $\xi(k)$ is Σ_1-conservative over $T + \neg\eta(k)$,
(v) $\{\xi(k) : k \in N\} \equiv \{\chi_k : k \in N\}$.

Proof. We combine the ideas of the proofs of Lemma 5.8 and Theorem 11. By Lemma 17, there is a PR formula $\gamma(x,z)$ such that for all φ,
(1)   if $T \vdash \varphi$, then $T \vdash \neg \exists z \gamma(\varphi, z)$,
(2)   if $T \nvdash \varphi$, then $\exists z \gamma(\varphi, z)$ Π_1-conservative over $T + \varphi$.
Let $\delta(u)$ be an arbitrary PR formula. Let $\kappa(z,u,x,y)$ and $\nu(z,u,x,y)$ be Π_1 formulas and $\mu(z,u,x,y,v)$ a PR formula such that
(3) $T \vdash \kappa(z,u,x,y) \leftrightarrow \forall v \mu(z,u,x,y,v)$,
(4) $T \vdash \neg \nu(z,u,x,0)$,
(5) $T \vdash \kappa(\delta,u,k,y) \leftrightarrow \nu(\delta,u,k,y) \vee$
$$\forall v([\Sigma_1](\neg \eta_\delta(k) \wedge \xi_\delta(k)), v) \to \neg \mathrm{Prf}_T(\xi_\delta(k), v)),$$
(6) $T \vdash \nu(\delta,u,k,y+1) \leftrightarrow \forall v(\neg \mu(\delta,u,k+1,y,v) \to \exists z < \max\{u,v\} \gamma(\eta_\delta(k), z))$,
where
$$\xi_\delta(x) := \forall u(\delta(u) \to \kappa(\delta,u,x,u \dotdiv x)),$$
$$\eta_\delta(x) := \forall u(\delta(u) \to \nu(\delta,u,x,u \dotdiv x)).$$
As in the proof of Lemma 5.8, (5) implies that
(7) $T \vdash \xi_\delta(k) \leftrightarrow \eta_\delta(k) \vee \forall v([\Sigma_1](\neg \eta_\delta(k) \wedge \xi_\delta(k)), v) \to \neg \mathrm{Prf}_T(\xi_\delta(k), v))$.
Let
$$\eta'_\delta(x) := \forall u(\delta(u) \to \nu(\delta,u,x,(u \dotdiv (x+1)) + 1)).$$
Then, by (6),
(8) $T \vdash \eta'_\delta(k) \leftrightarrow \forall uv(\delta(u) \wedge \neg \mu(\delta,u,k+1,u \dotdiv (k+1), v) \to$
$$\exists z < \max\{u,v\} \gamma(\eta_\delta(k), z)).$$
Let
$$\chi_{\delta,k} := \exists z(\gamma(\eta_\delta(k), z) \wedge \forall uv \leqslant z \neg (\delta(u) \wedge \neg \mu(\delta,u,k+1,u \dotdiv (k+1), v))).$$
Then $\chi_{\delta,k}$ is Σ_1 and (cf. Lemma 1.3)
(9) $T \vdash \chi_{\delta,k} \leftrightarrow \exists z \gamma(\eta_\delta(k), z) \wedge \forall uv(\delta(u) \wedge \neg \mu(\delta,u,k+1,u \dotdiv (k+1), v) \to$
$$\exists z < \max\{u,v\} \gamma(\eta_\delta(k), z))$$
and so, by (8),
(10) $T \vdash \chi_{\delta,k} \leftrightarrow \exists z \gamma(\eta_\delta(k), z) \wedge \eta'_\delta(k)$.
We now show that
(11) $T \vdash \eta_\delta(k) \to \xi_\delta(k)$,
(12)   if $T \vdash \xi_\delta(k)$, then $T \vdash \eta_\delta(k)$,
(13) $T \vdash \xi_\delta(k+1) \to \eta'_\delta(k)$,
(14)   if $T \vdash \delta(u) \to u > k$, then $T \vdash \eta'_\delta(k) \leftrightarrow \eta_\delta(k)$,
(15)   if $T \vdash \delta(u) \to u > k$ and $T \vdash \eta_\delta(k)$, then $T \vdash \xi_\delta(k+1)$.
(11) follows from (7). (12) follows from (7) by the same argument as in the proof of Lemma 5.8. (13) follows by predicate logic from (3) and (8). (14) is obvious.

To prove (15), assume $T \vdash \delta(u) \to u > k$ and $T \vdash \eta_\delta(k)$. Then, by (14), $T \vdash \eta'_\delta(k)$. Also, by (1), $T \vdash \neg \exists z \gamma(\eta_\delta(k), z)$. By (8), it follows that

$$T \vdash \forall uv(\delta(u) \to \mu(\delta,u,k+1,u \dotdiv (k+1),v))$$

and so, by (3), $T \vdash \xi_\delta(k+1)$. This proves (15).

It can now be shown that

if $\exists u\delta(u)$ is true, then $T \nvdash \xi_\delta(0)$.

The proof of this from (4), (12), (15) is the same as that of (6) in the proof of Lemma 5.8.

As in the proof of Lemma 5.8 we can now find a PR formula $\delta'(x)$ such that $\exists u\delta'(u)$ is false and $T \nvdash \xi_{\delta'}(0)$. Let $\xi(x) := \xi_{\delta'}(x)$, $\eta(x) := \eta_{\delta'}(x)$, $\chi_k := \chi_{\delta',k}$.

The verification of (i)–(iv) is now straightforward or much the same as in the proof of Lemma 5.8; this is where (13) is needed.

To prove (v), we first note that $\{\xi(k): k \in N\} \leq \{\chi_k: k \in N\}$ follows from (10), (14), (11). Next suppose $T + \{\chi_k: k \in N\} \vdash \pi$. There is then an m such that $T + \chi_0 \vdash \chi_1 \wedge ... \wedge \chi_m \to \pi$. By (i) and (ii), $T \nvdash \eta(0)$. Hence, by (2), $\exists z\gamma(\eta(0),z)$ is Π_1-conservative over $T + \eta(0)$. But then, by (10) and (14), $T + \eta(0) \vdash \chi_1 \wedge ... \wedge \chi_m \to \pi$. By (10), (14), (ii), (iii), $T + \chi_1 \vdash \eta(0)$. It follows that $T + \chi_1 \wedge ... \wedge \chi_m \vdash \pi$. Continuing in this way we eventually get $T + \eta(m) \vdash \pi$ and so, by (iii), $T + \xi(m+1) \vdash \pi$. This shows that $\{\chi_k: k \in N\} \leq \{\xi(k): k \in N\}$ and so (v) is proved. ∎

Proof of Theorem 16. Let $\xi(x)$ and $\eta(x)$ be as in Lemma 24. Let $a = d(\{\xi(k): k \in N\})$. Then, by Lemma 24(i), $a > 0_T$. Also, by Lemma 24(v) and Lemma 13, a is Σ_1. By Lemma 24(ii), $d(\xi(k)) \leq d(\eta(k))$ for every k. That $d(\xi(k))$ doesn't cup to $d(\eta(k))$ now follows from Lemma 24(iv).

Suppose b is Π_1 and $b \leq a$. Then, by Lemma 24(ii) and (iii), $b \leq d(\xi(k))$, for some k, and $d(\eta(k)) \leq a$. Since $d(\xi(k))$ doesn't cup to $d(\eta(k))$, it follows that b doesn't cup to a. ∎

Note that if a is as in Theorem 16, then a does not cup to any Π_1 degree. Indeed, let b be Π_1 and $\geq a$. If a cups to b, there is a Π_1 degree $c \leq a$ which cups to b. But then c cups to a, contrary to assumption.

Finally, we prove Theorem 5 (and a bit more). We have already observed that $d(\neg\pi)$ is the p.c. of $d(\pi)$. Thus, every Π_1 degree has a p.c. It follows that, in terms of our classification of degrees, the following result is the best we can do.

Theorem 17. There is a Σ_1 degree which has no p.c.

This is a consequence of the following strengthening of Lemma 19.

Lemma 25. There is a sentence σ such that $\{b \geq d(\neg\sigma): b \text{ is } \Sigma_1\}$ has no g.l.b.

To prove this, we need another:

Lemma 26. Suppose $\{\pi_k: k \in N\}$ is r.e. and let $G = \{d(\pi_k): k \in N\}$. Suppose there is no finite subset H of G such that $\cap H$ is a lower bound of G. Then G has no g.l.b.

Proof. Let $X = \{\pi: T + \pi_k \vdash \pi$ for every $k\}$. X is not r.e. This can be seen as follows. Let $R(k,m)$ be a primitive recursive relation such that $Y = \{k: \forall m R(k,m)\}$ is not r.e. and let $\rho(x,y)$ be a PR binumeration of $R(k,m)$. We may assume that $Z = \{\pi_k: k \in N\}$ is primitive recursive; let $\zeta(x)$ be a PR binumeration of Z. Finally, let $\eta(x) :=$
$$\forall z(\neg \rho(x,z) \to \exists u \leq z(\zeta(u) \land \mathrm{Tr}_{\Pi_1}(u))).$$
It is sufficient to show that
(1) $Y = \{k: \eta(k) \in X\}$.

If $k \in Y$, then, clearly, $\eta(k) \in X$. Suppose $k \notin Y$. Let m be such that not $R(k,m)$. Then $T + \eta(k) \vdash \bigvee Z|m$. By assumption, there is an n such that $T + \pi_n \not\vdash \bigvee Z|m$ and so $\eta(k) \notin X$. Thus, (1) holds and so X is not r.e.

Suppose $d(A) \leq d(\pi_k)$ for every k. Then $\mathrm{Th}(A) \cap \Pi_1 \subseteq X$. Since X is not r.e., it follows that there is a $\pi \in X$ such that $A \not\vdash \pi$. But then $\pi \leq \pi_k$ for every k and $d(\pi) \not\leq d(A)$. Thus, $d(A)$ is not the g.l.b. of G. ∎

Proof of Lemma 25. From the proof of Theorem 11 it is clear that there are (primitive) recursive functions f and g such that if π is any Π_1 sentence, then $f(\pi)$ is a Π_1 sentence, $g(\pi)$ is a Σ_1 sentence, and if $T \not\vdash \pi$, then $T < T + f(\pi) \equiv T + g(\pi) \leq T + \pi$.

We now define π_k and σ_k as follows. Let π_0 be any Π_1 sentence not provable in T. Next suppose π_k has been defined and $T \not\vdash \pi_k$. Let ψ be a Π_1 sentence undecidable in $T + \neg\pi_k$. Then $T < T + \pi_k \lor \psi < T + \pi_k$. Let $\sigma_k := g(\pi_k \lor \psi)$ and $\pi_{k+1} := f(\pi_k \lor \psi)$. Then $T \not\vdash \pi_{k+1}$.

For every k,
(1) $\pi_{k+1} \leq \sigma_k < \pi_k$.
By Theorem 5.4(a), there is a sentence σ such that
(2) $T + \sigma$ is a Π_1-conservative extension of $T + \{\neg\pi_k: k \in N\}$.
By (1) and (2),
(3) $\neg\sigma \leq \sigma_k$.
Moreover
(4)   if b is Σ_1 and $b \geq d(\neg\sigma)$, there is a k such that $b \geq d(\pi_k)$.
For suppose $b = d(\chi) \geq d(\neg\sigma)$, where χ is Σ_1. Then $T + \chi \vdash \neg\sigma$, whence $T + \sigma \vdash \neg\chi$. But then, by (2), there is a k such that $T + \neg\pi_k \vdash \neg\chi$, whence $T + \chi \vdash \pi_k$ and so $b \geq d(\pi_k)$.

Let $G = \{d(\pi_k): k \in N\}$. If $\{b \geq d(\neg\sigma): b$ is $\Sigma_1\}$ has a g.l.b. c, then, by (1), (3), (4), c is the g.l.b. of G. But from (1) it follows that no $d(\pi_k)$ is a lower bound of G. Hence, by Lemma 26, G has no g.l.b. and so $\{b \geq d(\neg\sigma): b$ is $\Sigma_1\}$ has no g.l.b. ∎

Proof of Theorem 17. Let σ be as in Lemma 25. By Lemma 6, for all B, $(T+\sigma)\downarrow B \leq T$ iff $B \leq T + \chi$ for all Σ_1 sentences χ such that $T + \chi \vdash \neg\sigma$. But then the p.c. of $d(\sigma)$, if it had one, would also be the g.l.b. of $\{b \geq d(\neg\sigma): b$ is $\Sigma_1\}$. Thus, by Lemma 25, $d(\sigma)$ has no p.c. ∎

Every Σ_1 degree is the p.c. of some degree. It is an open problem if the converse of this is true. If it is, the Σ_1 degrees can be characterized in a purely algebraic way as those degrees that are p.c.s.

Exercises for Chapter 7

In the following Exercises we assume that PA ⊣ T and that A, B, etc. are extensions of T.

1. Suppose $G \subseteq D_T$. G is *independent* if for any disjoint finite subsets G_0 and G_1 of G, $\cap G_0 \not\leq \cup G_1$. ($\cap \emptyset = 1_T$, $\cup \emptyset = 0_T$.) (Thus, for example, $\emptyset$ is independent and $\{a\}$ is independent iff $0_T < a < 1_T$.) Show that for every finite independent set G, there are degrees b_0, b_1 such that $G \cup \{b_i\}$ is independent, $i = 0, 1$, and $b_0 \cap b_1 = 0_T$. Conclude that every finite independent set is included in $2^{\aleph_0}$ many maximal independent sets.

2. Suppose $a < b$.
    (a) Show that there is a $c \in (a,b)$ such that for every $d \geq a$, if $c \cup d = b$, then $d = b$.
    (b) Show that there is a $c \in (a,b)$ such that for every $d \leq b$, if $c \cap d = a$, then $d = a$.

3. Suppose $a < b$ and, if T is Σ_1-sound, $b < 1_T$. For $c \in [a,b]$, let c^* be the *complement of c in* $[a,b]$ if it exists, i.e., $c \cap c^* = a$ and $c \cup c^* = b$. (Complements are unique.) Let $\text{Cpl}_{a,b}$ be the set of degrees in $[a,b]$ having complements in $[a,b]$ and let $\text{CPL}_{a,b} = (\text{Cpl}_{a,b}, \cap, \cup, *, a, b)$. Then $\text{CPL}_{a,b}$ is a Boolean algebra.
    (a) Show that if $c, d \in \text{Cpl}_{a,b}$ and $c < d$, then $\text{Cpl}_{a,b} \cap (c,d) \neq \emptyset$. (It follows that the Boolean algebras $\text{CPL}_{a,b}$ are (denumerable and) atomless and therefore isomorphic.)

(b) Show that $Cpl_{a,b} \neq [a,b]$.
(c) Show that if $a \leq c < d \leq b$, then $[c,d] \not\subseteq Cpl_{a,b}$.

4. Suppose a is Σ_1.
   (a) Show that if $a < b < 1_T$, there is a degree c such that $0_T < c < b$ and $a \cap c = 0_T$.
   (b) Show that if $a < b$ and b is high, then $a \ll b$. Conclude that if $b_i > a$, $i = 0, 1$, and $b_0 \cap b_1 = a$, then b_0 and b_1 are low.

5. Show that for every low degree a, there is a low Π_1 degree $\geq a$.

6. Referring to the proof of Theorem 4, show that there is a primitive recursive function g such that ψ can be replaced by the sentence
$$\chi := \forall u (Prf_B(\bot,u) \rightarrow \exists z < g(u) Prf_T(\bot,z)),$$
similar to θ. [Hint: Define g in such a way that $PA \vdash \neg \varphi \rightarrow \chi$.]

7. (a) Show that there is an r.p. degree a which is not Π_1 (compare Lemma 11). [Hint: Let $\kappa(x)$ be as in Exercise 2.24 and let $a = d(\{\kappa(k) : k \in N\})$].
   (b) Improve (a) by showing that there is a non-Π_1 Σ_1 degree a which is r.p. (compare Exercise 14(c)).

8. Suppose $A \dashv B$. Show that there is a Δ_2 sentence φ such that $A + \varphi \simeq B$ (compare Corollary 6.10 and Theorem 8).

9. Show that there is a Σ_1 sentence σ such that $0_T < d(\sigma) < 1_T$ and for every Σ_1 sentence χ, if $\sigma \leq \chi$, then $T + \chi \vdash \sigma$.

10. (a) Show that if b is the p.c. of a, then $a \cup b \gg 0_T$. (It follows that for every π, $d(\pi) \cup d(\neg\pi) \ggg 0_T$).
    (b) Show that if a is Π_1 and high, there is a low Π_1 degree b with p.c. b^* such that $a = b \cup b^*$. [Hint: Let b be as in the proof of Theorem 15.]
    (c) Show that there is a degree $a \ggg 0_T$ such that every Π_1 degree $\leq a$ is low.

11. Show that there is a degree of the form $d(\sigma \vee \pi)$ which is neither Σ_1 nor Π_1.

12. Suppose $a < b < 1_T$. Show that
    (a) there is a degree $c < 1_T$ such that for every d, if $b \cap d = a$, then $d \leq c$,
    (b) there is a degree $c > 0_T$ such that for every d, if $a \cup d = b$, then $d \geq c$.

13. (a) Verify that in any distributive lattice, for any a, b, the intervals [a ∩ b, a] and [b, a ∪ b] are isomorphic.

(b) Show that there are degrees a, b, c, d such that a ≪ b, c < d, not c ≪ d, and [a,b] and [c,d] are isomorphic.

14. (a) Verify that in any distributive lattice, if a < b < c and [a,c] satisfies the r.p., so does [b,c].

(b) Show that for each degree $a < 1_T$, there is a b such that $a < b < 1_T$ and [a,b] does not satisfy the r.p

(c) The non-r.p. degree a defined in the proof of Lemma 12 is high (cf. Exercise 10(a)). Show that there is a Σ_1 degree which is not r.p. Conclude from Exercise 7(b) that there are non-Π_1 Σ_1 degrees such that $[0_T, a]$ and $[0_T, b]$ are not isomorphic. [Hint: Use Theorem 14'(a).]

15. (a) Suppose φ and X are as in Lemma 15. Show that if φ ⩽ X, then φ ≪ X.

(b) Suppose a < b. Show that there are c, d such that a ⩽ c < d ⩽ b and [c,d] contains no B_1 degree.

16. (a) Show that there are low cupping degrees.

(b) Show that there is a high (Π_1) degree a which is not cupping. [Hint: Let σ, π be such that T ⊬ σ and σ is Π_1-conservative over T, π is Σ_1-conservative over T + σ, and ¬π is Π_1-conservative over T + σ. Let a = d(¬σ∨π).]

(c) Show that there are degrees a_0, a_1 such that $0_T \ll a_0 \ll a_1 < 1_T$ and $a_0 \ll_\cup a_1$.

17. (a) Show that there is a Π_1 degree a such that for every Π_1 degree b, if $a \cap b = 0_T$, then a ∪ b is low.

(b) Show that there is a low (Π_1) degree a such that for every degree b, if $a \cap b = 0_T$, then a ∪ b is not cupping. [Hint: Let b be a high (Π_1) non-cupping degree (see Exercise 16(b)). Define a sentence σ such that d(σ) > 0_T and d(σ) ∪ d(¬σ) ⩽ b; use Theorem 11. Let a = d(¬σ).]

18. Show that there are degrees a, b such that a is Σ_1, b is both Σ_1 and Π_1, and a ∪ b is not B_1.

19. Prove Lemma 14 by letting θ_0 be a Π_1 Rosser sentence for T and $\theta_1 :=$
$$\forall u(\text{Prf}_T(\neg\theta_0, u) \to \exists z \leqslant u \text{Prf}_T(\theta_0, z)).$$
Conclude that $d(\theta_0)$ is Σ_1 (compare Exercise 6.9).

20. (a) Let G be the set of degrees a for which there is a smallest Σ_1 degree $\geq$ a (compare Lemma 18). Show that G is closed under $\cap$ and $\cup$.

(b) Let H be the set of high degrees. Show that there is a Π_1 degree not in $\mathrm{Cl}(H \cup \Sigma_1)$.

21. Suppose $a \in \mathrm{Cl}(\Sigma_1)$ and $a > 0_T$. Show that there is a degree $b < a$ such that $[b,a] \subseteq \mathrm{Cl}(\Sigma_1)$.

22. Show that for every degree $a > 0_T$, there is a degree $b \leq a$ such that $b \notin \mathrm{Cl}(B_1)$. [Hint: Let π and b be such that $0_T < d(\pi) \leq a$, $d(\pi)$ is low, $d(\neg\pi) \ll b$, not $0_T \lll b$. Then $b \cap d(\pi) \notin \mathrm{Cl}(B_1)$.]

23. (a) Show that not all non-Π_1 Σ_1 degrees are as stated in Theorem 16.

(b) Improve Theorem 16 by showing that for every degree $b > 0_T$, there is a Σ_1 degree a such that $0_T < a < b$ and no Π_1 degree cups to a.

24. Show that in contrast to Lemma 26 we have the following: There is a set $G = \{d(\sigma_k): k \in N\}$ of Σ_1 degrees, where $\{\sigma_k: k \in N\}$ is (primitive) recursive, such that $\cap H > 0_T$ for every finite subset H of G and $\cap G = 0_T$. [Hint: Let a be high and such that there is no high Π_1 degree $\leq$ a (cf. Exercise 10(c)). Let $A \in a$ and let $\sigma_k := \neg \mathrm{Con}_{A|k}$.]

25. (a) Show that the set $\{\sigma: d(\sigma) \text{ is } \Pi_1\}$ is complete Σ_3^0. [Hint: The set $\{e: W_e$ is cofinite$\}$ is complete Σ_3^0. (W_e is the r.e. set with index e.)]

(b) Show that the set $\{\sigma: d(\sigma) \text{ has a p.c.}\}$ is complete Σ_3^0.

26. (a) Show that there is a PR formula $\delta(u)$ such that if θ is defined as in the proof of Theorem 11, then $d(\neg\theta)$ isn't Π_1. (For a proof of this, see Lindström (1993).)

(b) Let θ be as in (a). Show that $d(\neg\theta)$ has a p.c. Conclude that there is a non-Π_1 Σ_1 degree which has a p.c.

Notes for Chapter 7

The lattice $\mathbf{D}_T$ was introduced by Lindström (1979), (1984b); a related lattice $\mathbf{V}_T$ (degrees of finite extensions of T) has been defined by Švejdar (1978) (see also Jeroslow (1971a)). (By Theorem 6.11(a), $\mathbf{V}_T$ and $\mathbf{D}_T$ are isomorphic.) A quite different lattice of degrees (types) of (local) interpretability of not necessarily arithmetical theories has been introduced by

Mycielski (1977); cf. also Mycielski, Pudlák, Stern (1990). Theorem 1 is due to Lindström (1979), (1984b) and (for V_T) to Švejdar (1978). Corollary 1 is, modulo Theorem 6.6, a restatement of the equivalence of Exercise 2.25(i) and (ii). The proof of Theorem 4 was suggested by the proof of a related result in Hájková II (1971). Theorem 7 is new; the term "reduction principle" is borrowed from descriptive set theory and recursion theory (cf. Soare (1987)). (The only way of showing that intervals are isomorphic known so far is given in Exercise 13(a) and works in all distributive lattices.) The remaining results of §1 are due to Lindström (1979), (1984b). In connection with the proof of Theorem 4, see Exercise 6. Lemmas 11 and 12 lead to the question if there is a non-Π_1 r.p. degree; this question is answered in Exercise 7.

Theorem 8 (with a slightly different proof; see Exercise 6.12(a)) is due to Montagna (cf. Lindström (1993)). Theorem 9 is due to Lindström (1979), (1984b), (1993); (a) and (c) were also proved by Švejdar (1978).

Theorem 10 is due to Lindström (1979), (1984b); (a) and the first half of (b) were also proved by Švejdar (1978). Theorems 14 and 16 are new, they were announced in Lindström (1993), where a weaker form of Theorem 16 is proved; Theorem 16 leads to the question if there is a Σ_1 degree a such that no Π_1 degree caps to a; this is answered negatively in Exercise 5; in connection with Theorem 16, see also Exercise 23. The remaining results of §3 are due to Lindström (1984b), (1993), (1997). The definition of the sentences φ_n and ψ_n in the proof of Theorem 14(a) and the observations concerning these sentences, except (8), were first used by Misercque (1985), (1985a) in a different context. Theorem 17 leads to the question if no non-Π_1 Σ_1 degree has a p.c.; this question is answered in Exercise 26(b).

8. GENERALIZATIONS

So far our results have been explicitly stated (and proved) only for theories of first order arithmetic. But, as mentioned in the introduction, they hold, after suitable reformulation, in a much more general setting. Needless to say, we are not going to show this in every detail. In fact, we shall skip Chapters 3, 5, 7 altogether and concentrate on some of the main results of Chapters 2, 4, and 6. These examples should enable the reader to generalize (most of) the results of the preceding chapters.

In this chapter the theories S, T, etc. are no longer arithmetical theories, but they are still consistent and primitive recursive and we assume that the languages of these theories are always finite. L_T is the language of T. T is a *pure* extension of S if S ⊣ T and $L_T = L_S$. Lower case Greek letters are now used for formulas of L_T as well as for formulas of L_A.

We assume that the reader can extend the definition of t: S ≤ T to the present more general setting. Let $t^{-1}(T) = \{\varphi: T \vdash t(\varphi)\}$. Then $t^{-1}(T) \vdash \psi$ iff $T \vdash t(\psi)$. Since L_S is finite, t is primitive recursive.

The following lemma is immediate.

Lemma 1. (a) t: S ≤ T iff S ⊣ $t^{-1}(T)$.
(b) t: $t^{-1}(T)$ ≤ T and so $t^{-1}(T)$ ≤ T; in fact, t: $t^{-1}(T) \trianglelefteq T$; it follows that $t^{-1}(T)$ is consistent.
(c) $t^{-1}(T + t(\varphi))$ ⊣⊢ $t^{-1}(T) + \varphi$.

§1. Incompleteness. Our first result, Gödel's incompleteness theorem, is a straightforward generalization of Theorem 2.1; $\delta_t(x,y)$ is a formula defining t as in Fact 2.

Theorem 1. Suppose t: Q ≤ T. Let φ be such that
(Gt) $Q \vdash \varphi \leftrightarrow \neg \exists y(\delta_t(\varphi,y) \wedge \Pr_T(y))$.
Then φ is a true Π_1 sentence such that $T \nvdash t(\varphi)$. Hence if $t^{-1}(T)$ is Σ_1-sound, then also $T \nvdash \neg t(\varphi)$.

By Theorem 1, for each t: Q ≤ T, there is a true Π_1 sentence φ_t such that $T \nvdash t(\varphi_t)$. By a similar generalization of Rosser's theorem, we obtain a Π_1 sentence θ_t such that $T \nvdash t(\theta_t)$ and $T \nvdash \neg t(\theta_t)$. This result can be improved by showing that there is a single Π_1 sentence ψ such that $T \nvdash t(\psi)$ and $T \nvdash \neg t(\psi)$ for every t: $Q \leq T$:

Theorem 2. There is a (true) Π_1 sentence ψ, such that $Q + \psi \nleq T$ and $Q + \neg \psi \nleq T$.

Proof. $\{\varphi: Q + \varphi \leq T\}$ is r.e. (Lemma 6.5) and monoconsistent with Q. Now use Lemma 2.1. ∎

Our next result, Gödel's second incompleteness theorem, is a generalization of Theorem 2.4(a). Since each t is primitive recursive, in PA we may use t as a function symbol.

Theorem 3. Suppose t: PA $\leq$ T.
 (a) $T \nvdash t(\mathrm{Con}_T)$.
 (b) If $\tau(x)$ is any Σ_1 numeration of T, then $T \nvdash t(\mathrm{Con}_\tau)$.

Proof. We prove (a); the proof of (b) is almost the same. Let φ be as in the proof of Theorem 1. Then $\mathrm{PA} \vdash \neg\varphi \to \mathrm{Pr}_T(t(\varphi))$. Moreover, $\neg\varphi$ being Σ_1, $\mathrm{PA} \vdash \neg\varphi \to \mathrm{Pr}_Q(\neg\varphi)$. Since Q is finite, it follows, by Fact 12 (Chapter 6), that $\mathrm{PA} \vdash \mathrm{Pr}_Q(\neg\varphi) \to \mathrm{Pr}_T(\neg t(\varphi))$. We may now conclude that $\mathrm{PA} \vdash \neg\varphi \to \neg\mathrm{Con}_T$ and so $T \vdash t(\mathrm{Con}_T) \to t(\varphi)$. But then, by Theorem 1, $T \nvdash t(\mathrm{Con}_T)$, as desired. ∎

We shall say that t is a *reflexive* interpretation of S in T, t: $S \leq^r T$, if t: $S \leq T$ and for every k, $T \vdash t(\mathrm{Con}_{T|k})$. S is *reflexively interpretable*, $S \leq^r T$, if there is a t such that t: $S \leq^r T$. t is an *essentially reflexive* interpretation of S in T, t: $S \leq^{er} T$, if t: $S \leq^r T'$ for every pure extension T' of T. S is *essentially reflexively interpretable*, $S \leq^{er} T$, if there is a t such that t: $S \leq^{er} T$.

As in Chapter 2, Theorem 3 has the following:

Corollary 1. If $\mathrm{PA} \leq^r T$, then T is not finitely axiomatizable.

$L_{ST} = \{\in\}$ is the language of (first order) set theory. ZF is Zermelo-Fraenkel set theory. Let t_s be the standard interpretation of arithmetic in set theory.

Fact 13. t_s: PA $\leq^{er}$ ZF.

Combining this with Corollary 1, we get (compare Corollary 2.1):

Corollary 2. No consistent pure extension of ZF is finitely axiomatizable.

This result will be strengthened §§2 and 3 (Corollaries 3 and 7).

A nonreflexive interpretation of PA in ZF can be defined as follows. The theory $t_s^{-1}(ZF)$ is Σ_1-sound. Hence, by Corollary 6.9(b), there is a faithful interpretation t': PA $\trianglelefteq t_s^{-1}(ZF)$. Let $t = t_s t'$. Since t_s: $t_s^{-1}(ZF) \trianglelefteq ZF$ (Lemma 1(b)), it follows that t: PA $\trianglelefteq$ ZF. There is a finite subtheory ZF|k of ZF such that PA $\vdash \text{Con}_{ZF|k} \to \text{Con}_{PA}$. (This follows from the fact that t_s is a "natural" interpretation of PA in a finite subtheory of ZF.) Since PA $\nvdash \text{Con}_{PA}$, it follows that PA $\nvdash \text{Con}_{ZF|k}$. Since t is faithful, this implies that ZF $\nvdash t(\text{Con}_{ZF|k})$ and so t is not reflexive.

§2. Axiomatizations. In this § we shall restrict ourselves to generalizing Theorem 4.2. We shall need the following generalization of part (a) of the Fixed Point Lemma; the proof is left to the reader.

Lemma 2. Suppose t: PA $\leqslant$ T and let $v_n(x) := t(x = n)$. Let $\gamma(x)$ be any formula of L_T. There is then a sentence φ such that
$$T \vdash \varphi \leftrightarrow \exists y (v_\varphi(y) \wedge \gamma(y)).$$

We assume given a hierarchy $H = (H_0, H_1, \ldots)$ of the formulas of L_T satisfying closure conditions similar to those satisfied by the hierarchy $(\Sigma_0, \Sigma_1, \ldots)$. Thus, each H_k is a primitive recursive set of formulas, $H_k \subseteq H_{k+1}$, and $\cup \{H_k : k \in N\}$ is the set of all formulas of T. Let $H_k(x)$ be a PR binumeration of H_k.

We assume that for each k, there is an H_k *partial truth-definition* for H_k in T i.e. an H_k formula $\text{Tr}_k(x)$ such that for every H_k sentence φ,
(Tr_k) $T \vdash \varphi \leftrightarrow \exists x (v_\varphi(x) \wedge \text{Tr}_k(x))$.
We also assume that the formulas $\text{Tr}_k(x)$ are *mutatis mutandis* as in Fact 10(a).

A set X of sentences of T is said to be H-*bounded* if there is a k such that $X \subseteq H_k$. Let
$$\text{RFN}_S^t = \{\forall x (t(H_k(x)) \wedge t(\text{Pr}_S(x))) \to \text{Tr}_k(x)) : k \in N\}.$$

Theorem 4. Suppose t: PA $\leqslant$ T, $t(\Sigma_1) \subseteq H_0$, and $T \vdash \text{RFN}_S^t$. If X is any H-bounded set of sentences such that $T \dashv S + X$, then $S + X$ is inconsistent.

Proof. This proof is essentially the same as the proof of Theorem 4.2. Let n be such that $X \subseteq H_n$. Let ψ be such that

§3. *Interpretability* 149

(1) $\quad T \vdash \psi \leftrightarrow \exists y(v_\psi(y) \wedge \forall xz(t(H_n(x)) \wedge Tr_n(x) \wedge t(z = (x \to y)) \to \neg t(Pr_S(z))))$.

By assumption, we have

(2) $\quad T \vdash \forall y(v_\psi(y) \to \forall xz(t(H_n(x)) \wedge t(z = (x \to y)) \wedge t(Pr_S(z)) \to (Tr_n(x) \to \psi)))$.

(1) and (2) imply that

(3) $\quad T \vdash \psi$.

Suppose $T \dashv S + X$. By (3), there is then a conjunction θ of members of X such that $S + \theta \vdash \psi$. It follows that

(4) $\quad T \vdash \exists z(v_{\theta \to \psi}(z) \wedge t(Pr_S(z)))$.

Also, by (1) and (3),

$\quad S + X \vdash \neg \exists z(v_{\theta \to \psi}(z) \wedge t(Pr_S(z)))$.

But then, by (4), $S + X$ is inconsistent. ∎

Theorem 4 can be applied to set theory. We define Σ_k^{ST} and Π_k^{ST} as follows. Let $\Sigma_0^{ST} = \Pi_0^{ST}$ be the set of formulas of L_{ST} all of whose quantifiers are *bounded*, i.e. of the form $\exists x \in y$ or $\forall x \in y$. Σ_{k+1}^{ST} and Π_{k+1}^{ST} are then the least sets closed under bounded quantification such that $\Sigma_k^{ST} \subseteq \Pi_{k+1}^{ST}$, and $\Pi_k^{ST} \subseteq \Sigma_{k+1}^{ST}$, Σ_{k+1}^{ST} is closed under existential quantification and Π_{k+1}^{ST} is closed under universal quantification. A set X of formulas of L_{ST} is *bounded* if $X \subseteq \Sigma_k^{ST}$ for some k. We then have the following:

Fact 14. The assumptions of Theorem 4 are satisfied when $t = t_s$, $H_k = \Sigma_{k+1}^{ST}$, $T = ZF$, and $S = \emptyset$.

From Theorem 4 and Fact 14, we get:

Corollary 3. There is no bounded and consistent set X of sentences of L_{ST} such that $ZF \dashv X$.

§3. Interpretability. In this § we show that the relevant results of Chapter 6 generalize quite easily to the present more general setting. In using results from Chapter 6 we shall take advantage of the fact that in these results the theories S, S_0, etc. need not be formalized in L_A.

Theorem 5. If $t: PA \leqslant^r T$, then $T \leqslant t^{-1}(T)$ and so $t^{-1}(T) \equiv T$.

Proof. By assumption, $T \vdash t(Con_{T|k})$ for every k. It follows that

$t^{-1}(T) \vdash \text{Con}_{T|k}$ for every k. But then, by Lemma 6.2, $T \leqslant t^{-1}(T)$. ∎

Let us say that $t: PA \leqslant T$ is *optimal with respect to Γ sentences* if for every $t': PA \leqslant T$ and every Γ sentence φ, if $T \vdash t'(\varphi)$, then $T \vdash t(\varphi)$.

Suppose $t: PA \leqslant T$. There is then a Σ_1 (true Π_2) sentence φ such that $t^{-1}(T) + \varphi \leqslant t^{-1}(T)$ and $t^{-1}(T) \nvdash \varphi$ (cf. Theorem 6.9 and the proof of Theorem 6.10). Let $t': t^{-1}(T) + \varphi \leqslant t^{-1}(T)$ and set $t'' = tt'$. Then $t'': PA + \varphi \leqslant T$. Since $T \nvdash t(\varphi)$, it follows that t is not optimal with respect to Σ_1 (true Π_2) sentences. (If $t^{-1}(T)$ and φ are true, we can also achieve that $t''^{-1}(T)$ is true, since, by Corollary 6.9(b), $t^{-1}(T) + \varphi \trianglelefteq t^{-1}(T)$.) In contrast to this we have the following:

Corollary 4. If $t: PA \leqslant^r T$, then t is optimal with respect to Π_1 sentences.

Proof. Suppose $t': PA \leqslant T$. $t'^{-1}(T) \leqslant T$ and, by Theorem 5, $T \leqslant t^{-1}(T)$. Thus, $t'^{-1}(T) \leqslant t^{-1}(T)$. But then, by Theorem 6.6, $t'^{-1}(T) \dashv_{\Pi_1} t^{-1}(T)$. ∎

Since $PA + \text{Con}_{PA} \leqslant ZF$, the nonreflexive $t: PA \leqslant ZF$ defined at the end of §1 is not optimal with respect to Π_1 sentences.

From Fact 13 (this chapter) and Corollary 4 we get:

Corollary 5. If T is a pure extension of ZF, then $t_s: PA \leqslant T$ is optimal with respect to Π_1 sentences.

Corollary 5 can also be proved directly in the following way. Let T be as assumed. For any $t: PA \leqslant T$ and any model **M** of T, let **M**^{t} be the model of PA defined in **M** by t. In **M**^{t_s} induction holds for every formula of L_{ST}. It follows that if $t: PA \leqslant T$, then **M**^{t_s} is isomorphic to an initial segment of **M**^{t} (compare the proof of Theorem 6.7) and so every Π_1 sentence true in **M**^{t} is true in **M**^{t_s}.

From Lemma 1(b) and Theorem 6.1, we get:

Lemma 3. There is a Σ_1 numeration $\tau'(x)$ of $t^{-1}(T)$ such that $PA \vdash \text{Con}_\tau \rightarrow \text{Con}_{\tau'}$.

Theorem 6.2 can now be generalized as follows:

Theorem 6. Suppose $t: PA \leqslant^r T$. Then $T + t(\text{Con}_T) \not\leqslant T$.

Proof. Suppose $T + t(\text{Con}_T) \leqslant T$. Then, by Lemma 1(b) and (c), and Theorem 5,

$t^{-1}(T) + \mathrm{Con}_\tau \leqslant T \leqslant t^{-1}(T)$. Let $\tau'(x)$ be as in Lemma 3. It follows that $t^{-1}(T) + \mathrm{Con}_{\tau'} \leqslant t^{-1}(T)$, contradicting Theorem 6.2. ∎

Corollary 6. If T is a pure extension of ZF, then $T + t_s(\mathrm{Con}_T) \not\leqslant T$.

Theorem 6.3 has the following generalization:

Theorem 7. Suppose $PA \leqslant^r T$. Then T is not interpretable in any finite subtheory of T.

Proof. Suppose $T \leqslant T|m$. Let t be such that $t: PA \leqslant^r T$. By Theorem 6.1, there is a Σ_1 numeration $\tau(x)$ of T such that $PA \vdash \mathrm{Con}_{T|m} \to \mathrm{Con}_\tau$. It follows that $T \vdash t(\mathrm{Con}_{T|m}) \to t(\mathrm{Con}_\tau)$. Also, by assumption, $T \vdash t(\mathrm{Con}_{T|m})$. But then $T \vdash t(\mathrm{Con}_\tau)$, contradicting Theorem 3(b). ∎

Corollary 7. If T is a pure extension of ZF, then T is not interpretable in any finite subtheory of T.

The Orey–Hájek lemma in the present setting reads as follows:

Lemma 4. Suppose $t: PA \leqslant^r T$. Then $S \leqslant T$ iff $T \vdash t(\mathrm{Con}_{S|k})$ for every k.

Proof. By Lemma 6.2, $S \leqslant t^{-1}(T)$ iff $t^{-1}(T) \vdash \mathrm{Con}_{S|k}$ for every k. ∎

As in Chapter 6 we get, from Lemma 4, the following version of Orey's compactness theorem.

Theorem 8. Suppose $PA \leqslant^r T$. Then $S \leqslant T$ iff for every k, $S|k \leqslant T$.

Theorems 5 and 6.6 imply the following generalization of Theorem 6.6; we use A, B for pure extensions of T.

Theorem 9. Suppose $t: PA \leqslant^{er} T$. Then $A \leqslant B$ iff $t^{-1}(A) \leqslant t^{-1}(B)$ iff $t^{-1}(A) \dashv_{\Pi_1} t^{-1}(B)$.

We conclude by generalizing Theorems 6.8 and 6.9; the generalization of Theorem 6.10 is left to the reader.

Theorem 10. If $t: PA \leqslant^{er} T$, then $T + \neg t(\mathrm{Con}_T) \leqslant T$.

Proof. Let $\tau'(x)$ be as in Lemma 3. By Theorem 6.8, $t^{-1}(T) + \neg\text{Con}_{\tau'} \leqslant t^{-1}(T)$. It follows that $t^{-1}(T) + \neg\text{Con}_T \leqslant t^{-1}(T)$. But then, by Lemma 1(c) and Theorem 5, $T + \neg t(\text{Con}_T) \leqslant t^{-1}(T + \neg t(\text{Con}_T)) \leqslant t^{-1}(T) + \neg\text{Con}_T \leqslant t^{-1}(T) \leqslant T$ and so $T + \neg t(\text{Con}_T) \leqslant T$, as desired. ∎

Theorem 11. Suppose $t: \text{PA} \leqslant^{\text{er}} T$ and X is r.e. and monoconsistent with T. There is then a sentence φ such that $T + \varphi \leqslant T$ and $\varphi \notin X$; φ can be taken to be of the form $t(\psi)$, where ψ is Σ_1.

Proof. Let $Y = \{\theta: t(\theta) \in X\}$. Then Y is r.e. and monoconsistent with $t^{-1}(T)$. By Theorem 6.9, there is a Σ_1 sentence ψ such that $t^{-1}(T) + \psi \leqslant t^{-1}(T)$ and $\psi \notin Y$. By Lemma 1(c) and Theorem 5, $T + t(\psi) \leqslant T$. Clearly, $t(\psi) \notin X$. ∎

Theorem 11 has the following application, where GB is Gödel-Bernays (finite) set theory (compare Corollary 6.6).

Corollary 8. There is a Σ_1 sentence φ such that $\text{ZF} + t_s(\varphi) \leqslant \text{ZF}$ and $\text{GB} + t_s(\varphi) \not\leqslant \text{GB}$.

Notes for Chapter 8

Theorems 1 and 3 are, of course, (essentially) due to Gödel (1931), (1934) (cf. also Feferman (1960)). Theorem 2 is due to Montague (1957), (1962) (compare Exercise 6.1(a)). For a definition of the standard interpretation t_s of arithmetic (theory of finite ordinals) in set theory, see, for example, Mendelson (1987). Fact 13 and Corollary 2 are due to Montague (1961). Fact 14 is due to Lévy (1965). Theorem 4 and Corollary 3 are due to Kreisel and Lévy (1968), improving earlier work of Montague (1961). Corollaries 2 and 3 are given here as examples of applications of the corresponding general results; for a detailed discussion of similar, more general, and stronger results, see Kreisel and Lévy (1968). Lemma 4, Theorems 6, 7, 8, 9, 10, 11, and Corollary 8 are straightforward generalizations of the corresponding results in Chapter 6. The question if there is a sentence φ such that $\text{GB} + \varphi \leqslant \text{GB}$ and $\text{ZF} + \varphi \not\leqslant \text{ZF}$, raised in Hájek (1971), was answered affirmatively by Solovay; for this and related results, cf. Hájek and Pudlák (1993).

REFERENCES

Only works mentioned in the text (Notes) have been included among the references; for a more comprehensive bibliography, see Hájek and Pudlák (1993).

Artemov, S. N. (1979). Extensions of arithmetic and modal logics, Thesis, Steklov Math. Inst., Moscow, 1979. (Russian)

Beklemishev, L. D. (1995). Iterated local reflection versus iterated consistency, Annals of Pure and Applied Logic 75, 25–48.

Beklemishev, L. D. (1997). Notes on local reflection principles, Theoria 63, 139–146.

Bennet, C. (1986). On some orderings of extensions of arithmetic, Thesis, Dept. of Philosophy, Univ. of Göteborg.

Bennet, C. (1986a). Lindenbaum algebras and partial conservativity, Proc. Amer. Math. Soc. 97, 323–327.

Berarducci, A. (1990). The interpretability logic of Peano arithmetic, J. Symb. Logic 55, 1059–1089.

Boolos, G. (1979). The Unprovability of Consistency, Cambridge University Press, Cambridge, USA.

Boolos, G. (1979a). Reflection principles and iterated consistency assertions, J. Symb. Logic 44, 33–35.

Boolos, G. (1993). The Logic of Provability, Cambridge University Press, Cambridge, USA.

Carnap, R. (1934). Logische Syntax der Sprache, Springer.

Craig, W. (1953). On axiomatizability within a system, J. Symb. Logic 18, 30–32.

Davis, M. (ed.) (1965). The Undecidable, Raven Press.

Dzhaparidze, G. (see also Japaridze, G.) (1993). A generalized notion of weak interpretability and the corresponding modal logic, Annals of Pure and Applied Logic 61, 113–160.

Ehrenfeucht, A. and Feferman, S. (1960). Representability of recursively enumerable sets in formal theories, Arch. Math. Logic 5, 37–41.

Ehrenfeucht, A. and Mycielski, J. (1971). Abbreviating proofs by adding new axioms, Bull. Amer. Math. Soc. 77, 366–367.

Feferman, S. (1960). Arithmetization of metamathematics in a general setting, Fund. Math. 49, 35–92.

Feferman, S. (1962). Transfinite recursive progressions of axiomatic theories, J. Symb. Logic 27, 259–316.

Feferman, S. (1997). My route to arithmetization, Theoria 63, 168–181.
Feferman, S., Kreisel, G., Orey, S. (1960). 1-consistency and faithful interpretations, Arch. Math. Logic 6, 52–63.
Friedman, H. (1975). One hundred and two problems in mathematical logic, J. Symb. Logic 40, 113–129.
Gödel, K. (1931). Über formal unentscheidbare Sätze der Principia Mathematica und verwandter Systeme I, Monatsh. Math. Physik. 38, 173–198 (English translation in Davis (1965) and in van Heijenoort (1967), reprinted and English translation in Gödel (1986)).
Gödel, K. (1934). On undecidable propositions of formal mathematical systems (mimeographed lecture notes by S. C. Kleene and J. B. Rosser), Princeton (reprinted in Davis (1965) and in Gödel (1986)).
Gödel, K. (1936). Über die Länge von Beweisen, Ergebnisse eines math. Koll., 7, 23–24 (English translation in Davis (1965), reprinted and English translation in Gödel (1986)).
Gödel, K. (1986). Collected Works, vol. I: Publications 1929–1936, edited by S. Feferman et al., Oxford University Press.
Goryachev, S. N. (1986). On the interpretability of some extensions of arithmetic, Mat. Zametki 40, 561–572; English transl. in Math. Notes 40.
Guaspari, D. (1979). Partially conservative extensions of arithmetic, Trans. Amer. Math. Soc. 254, 47–68.
Hájek, P. (1971). On interpretability in set theories, Comment. Math. Univ. Carol. 12, 73–79.
Hájek, P. (1979). On partially conservative extensions of arithmetic, in: Logic Colloquium '78 (eds. M. Boffa et al.), North-Holland, Amsterdam, 225–234.
Hájek, P. (1984). On a new notion of partial conservativity, in: Computation and Proof Theory, Springer Lecture Notes in Math. 1104, 217–232.
Hájek, P. and Montagna, F. (1990). ILM is the logic of Π_1-conservativity, Arch. Math. Log. 30, 113–123.
Hájek P., Montagna F., Pudlák P. (1992). Abbreviating proofs using metamathematical rules, in: Proof Theory and Computational Complexity (eds. P. Clote and J. Krajíček), Oxford University Press, 197–221.
Hájek, P. and Pudlák, P. (1993). Metamathematics of First-Order Arithmetic, Perspectives in Mathematical Logic, Springer-Verlag.
Hájková, M. (1971). The lattice of bi-numerations of arithmetic I, II, Comment. Math. Univ. Carol. 12, 81–104, 281–306.

Hájková, M. and Hájek, P. (1972). On interpretability in theories containing arithmetic, Fund. Math. 76, 131–137.

van Heijenoort, J. (ed.) (1967). From Frege to Gödel, Harvard University Press.

Hilbert, D. and Bernays, P. (1934, 1939). Grundlagen der Mathematik I, II, Springer-Verlag, Berlin.

Ignatiev, K. (1991). Partial conservativity and modal logic, ITLI prepublication series, X–91–04, Institute for Language, Logic and Information, University of Amsterdam.

Japaridze, G. (see also Dzhaparidze, G.) (1994). A simple proof of the arithmetical completeness of Π_1-conservativity logic, Notre Dame JFL, 346–354.

Japaridze, G. and de Jongh, D. (1998). The logic of provability, in: Handbook of Proof Theory (ed. S. R. Buss), Elsevier Science B. V., 475–546.

Jensen, D. and Ehrenfeucht, A. (1976). Some problems in elementary arithmetics, Fund. Math. 92, 223–245.

Jeroslow, R. G. (1971a). Consistency statements in formal theories, Fund. Math. 72, 17–40.

Jeroslow, R. G. (1971b). Non-effectiveness in S. Orey's compactness theorem, Zeitschr. Math. Log. Grundl. Math. 17, 285–289.

de Jongh, D. and Montagna, F. (1989). Much shorter proofs, Zeitschr. Math. Log. Grundl. Math. 35, 247–260.

de Jongh, D. and Veltman, F. (1990). Provability logics for relative interpretability, in: Mathematical Logic (ed. P. P. Petkov), Plenum Press, 31–42.

Kaye, R. (1991). Models of Peano arithmetic, Oxford Logic Guides 15, Clarendon Press, Oxford.

Kent, C. F. (1973). The relation of A to Prov'A' in the Lindenbaum sentence algebra, J. Symb. Logic 38, 295–298.

Kleene, S. (1952a). Introduction to metamathematics, van Nostrand.

Kleene, S. (1952b). Finite axiomatizability of theories in the predicate calculus using additional predicate symbols, Memoirs Amer. Math. Soc. 10, 27–66.

Kotlarski, H. (1994). On the incompleteness theorems, J. Symb. Logic 59, 1414–1419.

Kotlarski, H. (1996). An addition to Rosser's theorem, J. Symb. Logic 61, 285–292.

Kotlarski, H. (1998). Other proofs of old results, Math. Log. Quart. 44, 474–480.

Kreisel, G. (1957). Independent recursive axiomatization, J. Symb. Logic 22, 109 (abstract).

Kreisel, G. (1962). On weak completeness of intuitionistic predicate logic, J. Symb. Logic 27, 139–158.

Kreisel, G. and Lévy, A. (1968). Reflection principles and their use for establishing the complexity of axiomatic systems, Zeitschr. für math. Logik 14, 97–142.

Kreisel, G. and Wang, H. (1955). Some applications of formalized consistency proofs, Fund. Math. 42, 101–110.

Kripke, S. (1963). "Flexible" predicates in formal number theory, Proc. Amer. Math. Soc. 13, 647–650.

Lévy, A. (1965). A hierarchy of formulas in set theory, Memoirs Amer. Math. Soc. 57.

Lindström, P. (1979). Some results on interpretability, in: Proceedings of the 5th Scandinavian Logic Symposium 1979, Aalborg University Press, Aalborg, 329–361.

Lindström, P. (1984a). On partially conservative sentences and interpretability, Proc. Amer. Math. Soc. 91, 436–443.

Lindström, P. (1984b). On certain lattices of degrees of interpretability, Notre Dame J. Formal Logic 25, 127–140.

Lindström, P. (1984c). On faithful interpretability, in: Computation and Proof Theory, Springer Lecture Notes in Math. 1104, 279–288.

Lindström, P. (1988). Partially generic formulas in arithmetic, Notre Dame J. Formal Logic 29, 185–192.

Lindström, P. (1993). On Σ_1 and Π_1 sentences and degrees of interpretability, Annals of Pure and Applied Logic 61, 175–193.

Lindström, P. (1996). Provability logic–a short introduction, Theoria 62, 19–61.

Lindström, P. (1997). Interpretability in reflexive theories–a survey, Theoria 63, 182–209.

Löb, M. (1955). Solution of a problem of Leon Henkin, J. Symb. Logic 20, 115–118.

Mendelson, E. (1987). Introduction to Mathematical Logic (3rd ed.), van Nostrand.

Misercque, D. (1983). Answer to a problem by D. Guaspari, in: Open days in Model Theory and Set Theory (eds. W Guzicki et al.), Poland, 181–183.

Misercque, D. (1985). Branches of the E-tree which are not isomorphic, Bulletin of the London Math. Soc. 17, 513–517.

Montagna, F. (1982). Relatively precomplete numerations and arithmetic, J. Phil. Logic 11, 419–430.

Montagna, F. (1992). Polynomially and superexponentially shorter proofs in fragments of arithmetic, J. Symb. Logic 57, 844–863.

Montagna, F. and Sorbi, A. (1985). Universal recursion theoretic properties of r.e. preordered structures, J. Symb. Logic 50, 397–406.

Montague, R. (1957). Two theorems on relative interpretability, Summer Institute for Symbolic Logic, Cornell University, 263–264.

Montague, R. (1961). Semantical closure and non-finite axiomatizability I, in: Infinitistic Methods, Warsaw, 45–69.

Montague, R. (1962). Theories incomparable with respect to relative interpretability, J. Symb. Logic 27, 195–211.

Montague, R. (1963). Syntactical treatments of modality, with corollaries on reflection principles and finite axiomatizability, in: Proceedings of a Colloquium on Modal and Many-Valued Logics, Acta Philosophica Fennica, Helsinki, 153–167.

Montague, R. and Tarski, A. (1957). Independent recursive axiomatizability, Summer institute for Symbolic Logic, Cornell University, 270.

Mostowski, A. (1952a). On models of axiomatic systems, Fund. Math. 39, 133–158.

Mostowski, A. (1952b). Sentences undecidable in formalized arithmetic, North-Holland.

Mostowski, A. (1961). A generalization of the incompleteness theorem, Fund. Math. 49, 205–232.

Mycielski, J. (1977). A lattice of interpretability types, J. Symb. Logic 42, 297–305.

Mycielski, J., Pudlák, P., Stern, A. (1990). A Lattice of Chapters of Mathematics, Memoirs Amer. Math. Soc. 426.

Myhill, J. R. (1972). An absolutely independent set of Σ_1^0 sentences, Zeitschr. Math. Log. Grundl. Math. 18, 107–109.

Orey, S. (1961). Relative interpretations, Zeitschr. für math. Logik 7, 146–153.

di Paola, R. (1975). A theorem on shortening the length of proof in formal systems of arithmetic, J. Symb. Logic 40, 398–400.

Parikh, R. (1971). Existence and feasibility in arithmetic, J. Symb. Logic 36, 494–508.

Pour-El, M. (1968). Independent axiomatization and its relation to the hypersimple set, Zeitschr. für math. Logik 14, 449–456.

Pour-El, M. and Kripke, S. (1967). Deduction preserving recursive isomorphisms between theories, Fund. Math. 61, 141–163.

Pudlák, P. (1985). Cuts, consistency statements and interpretations, J. Symb. Logic 50, 423–441.

Putnam, H. and Smullyan, R. (1960). Exact separation of recursively enumerable sets within theories, Proc. Amer. Math. Soc. 11, 574–577.

Quinsey, J. (1980). Some problems in logic, Thesis, Oxford University.

Quinsey, J. (1981). Sets of Σ_k-conservative sentences are Π_2^0 complete, J. Symb. Logic 46, 442 (abstract).

Rabin, M. (1961). Non-standard models and independence of the induction axiom, in: Essays on the Foundations of Mathematics, Jerusalem, 287–299.

Rosser, B. (1936). Extensions of some theorems of Gödel and Church, J. Symb. Logic 1, 87–91 (reprinted in Davis (1965)).

Ryll-Nardzewski, C. (1952). The role of the axiom of induction in elementary arithmetic, Fund. Math. 39, 239–263.

Sambin, G. (1976). An effective fixed point theorem in intuitionistic diagonalizable algebras, Studia Logica 35, 345–361.

Scott, D. (1962). Algebras of sets binumerable in complete extensions of arithmetic, in: Recursive Function Theory (ed. J. Dekker), Providence, 117–121.

Shavrukov, V. Yu. (1988). The logic of relative interpretability over Peano arithmetic, Steklov Mathematical Institute, Moscow. (Russian)

Shavrukov, V. Yu. (1993). Subalgebras of diagonalizable algebras of theories containing arithmetic, Dissertationes Mathematicae CCCXXIII, Warsaw.

Shavrukov, V. Yu. (1997). Interpreting reflexive theories in finitely many axioms, Fund. Math. 152, 99–116.

Shepherdson, J. (1960). Representability of recursively enumerable sets in formal theories, Arch. Math. Logic 5, 119–127.

Simmons H. (1988). Large discrete parts of the E-tree, J. Symb. Logic 53, 980–984.

Smoryński, C. (1980). Calculating self-referential statements, Fund. Math. 109, 189–210.

Smoryński, C. (1981a). Calculating self-referential statements: Guaspari sentences of the first kind, J. Symb. Logic 46, 329–344.

Smoryński, C. (1981b). Fifty years of self-reference in arithmetic, Notre Dame J. Formal Logic 22, 357–374.
Smoryński, C. (1985). Self-Reference and Modal Logic, Springer-Verlag.
Smoryński, C. (1985a). Nonstandard models and related developments, in: Harvey Friedman's Research on the Foundations of Mathematics, North-Holland, 179–229.
Soare, R. (1987). Recursively Enumerable Sets and Degrees, Perspectives in Mathematical Logic, Springer-Verlag.
Solovay, R. (1976). Provability interpretations of modal logic, Israel JM 25, 287–304.
Solovay, R. (1985). Infinite fixed-point algebras, Proceedings of Symposia in Pure Mathematics 42, Providence, 473–486.
Strannegård, C. (1997). Arithmetical Realizations of Modal Formulas, Thesis, Dept. of Philosophy, University of Göteborg.
Strannegård, C. (1999). Interpretability over Peano arithmetic, J. Symb. Logic 64, 1407–1425.
Švejdar, V. (1978). Degrees of interpretability, Comment. Math. Univ. Carol. 19, 789–813.
Tarski, A. (1933). Der Wahrheitsbegriff in den formalisierten Sprachen, Studia Philosophica 1, 261–405.
Tarski, A., Mostowski, A. and Robinson, R. (1953). Undecidable Theories, North-Holland, Amsterdam.
Verbrugge, L. C. (1992). Feasible interpretability, in: Arithmetic, Proof Theory and Computational Complexity (eds. P. Clote and J. Krajíček), Oxford Univ. Press, 387–428.
Verbrugge, L. C. (1994). The complexity of feasible interpretability, in: Feasible Mathematics II (eds. P. Clote and J. Remmel), Birkhäuser, Boston, 429–447.
Visser, A. (1990). Interpretability logic, in: Mathematical Logic (ed. P. P. Petkov), Plenum Press, 195–209.
Wang, H. (1951). Arithmetical models of formal systems, Methodos 3, 217–232.

INDEX

arithmetical hierarchy   13
arithmetical interpretation   83
axiomatization   62

Bernays–Löb provability conditions   17
binumerate   9
binumeration   10
bounded (quantifier)   10, 149
bounded (set of sentences)   63, 148
B_n (formula, sentence)   13
B_n (degree) (see Φ (degree))

cap   121
cofinal   123
complement (in lattice)   24, 141
complete (theory)   26
complete Π_n^0, Σ_n^0 (set)   23
conservative (partially)   74
correctly numerates   51
Craig's theorem   12
cup   121
cupping   122

decidable (in T, formula, sentence)   10
decidable (theory)   19
define (formula defines function)   9
degree (of interpretability)   118
distributive (lattice)   23
dual (of Γ)   13
Δ_n (degree) (see Φ (degree))
Δ_n (formula, sentence)   13

effectively e. i. (over)   71
e.i.   62
essentially infinite (over)   62
essentially reflexive (theory)   22
essentially reflexive (interpretation)   147
essentially unbounded (over)   63
essentially undecidable (theory)   19
e.u.   63

faithful (interpretation)   106
faithfully interpretable   106

finite extension   62
fixed point   17
Fixed Point Lemma   17, 148
Φ (formula, sentence)   13
Φ (degree)   127

GB (Gödel–Bernays set theory)   152
g.l.b.   23
Gödel number(ing)   1, 11
Gödel's (first) incompleteness theorem   26, 146
Gödel's second incompleteness theorem   29, 147
Gödel–Tarski theorem   20
greatest lower bound   23
Γ (degree) (see Φ (degree))
Γ (formula, sentence)   13
Γ-conservative   74
Γ-conservative extension   78
Γ-sound   14
Γ-subtheory   76

H-bounded   148
Henkin complete   99
hereditarily Γ-conservative   90
high   130
hypersimple   73

i.a.   67
independent (formula)   35
independent (set of degrees)   141
independent (on X over T)   52
interpretable   97
interpretation   96, 97
interpretation (arithmetical)   83
interval   125
irredundant (over)   67
irredundantly axiomatizable   67
irredundantly Γ-axiomatizable (i. Γ-a.)   69

lattice   23
least upper bound   23
Liar paradox   20

Löb's theorem 32
low 130
l.u.b. 23

model for Π-conservativity 83
monoconsistent 28

N, **N** 7
numerate 9
numeration 10

Orey compactness theorem 102, 151
Orey–Hájek lemma 101, 151
ω-consistent 46

PA 7
partial truth-definition 20, 148
p.c. 24
Peano Arithmetic 7
positively prime 55
p.p. 55
PR (primitive recursive formula) 10
provably recursive 14
pseudocomplement 24
Π_n (degree) (see Φ (degree))
Π_n (formula sentence) (see Γ (formula sentence))
Π_n-conservative (see Γ-conservative)
Π_n-sound (see Γ-sound)

Q 7

rank, rank_n 84, 85
reduction principle 125
Robinson's Arithmetic 7
Rosser sentence 27
Rosser's theorem 26
r.p. 125

self-prover 90
Shepherdson–Smoryński fixed point theorem 61
Σ_n (degree) (see Φ (degree))
Σ_n (formula sentence) (see Γ (formula sentence))
Σ_1-complete 17
Σ_n-conservative (see Γ-conservative)
Σ_n-sound (see Γ-sound)

Tarski theorem 20
theory 7
translation 96
true (sentence, theory) 7
truth-definition 20
truth-preserving 88
type of independence 54

undecidable (sentence) 26
undecidable (theory) 19

ZF (Zermelo-Fraenkel set theory) 147

NOTATION

L_A 6
$0, S, +, \times$ 6
$:=$ 7
$\top, \bot$ 7
$T + \varphi, T + X$ 7
$\vdash, \dashv$ 7
$Th(T)$ 7
$N, \mathbf{N}$ 7
Q 7
$x \leq y, x < y$ 7
PA 8
$\delta_f(x_0,\ldots,x_n,y)$ 8
$\exists x \leq y, \forall x \leq y, \exists x < y, \forall x < y$ 9
PR 10
$\langle x,y \rangle, (x)_y, \langle k,m \rangle, (k)_m$ 10, 11
p_m 11
S, S_0, T, T', A, B etc. (conventions) 13
T (convention) 13
Σ_n, Π_n 13
$B_n, \Gamma, \Gamma^+, \Gamma^d$ 13
Φ, Φ^T 13
Δ_n 13
$Sbst_k$ 15
$Subst_k$ 15
$\xi(\dot{x}), \eta(\dot{x},\dot{y})$ 15
$Prf_\sigma(x,y)$ 15
$Pr_\sigma(x)$ 16
Con_σ 16
$\sigma|y, \sigma+y$ 16
$Prf_S(x,y), Pr_S(x), Con_S, Prf_{S+z}(x,y), Prf_{S|z}(x,y)$ 16
$\Gamma(x)$ 20

$Sat_\Gamma(x,y), Tr_\Gamma(x)$ 20
$Sat_{B_n}(x,y), Tr_{B_n}(x)$ 21
$X|k$ 22
Π_n^0, Σ_n^0 23
φ^i 28
$Ref(T)$ 40
$Prf^\mu(x,y)$ 47
X^c 52
$P, F, F(\langle \varphi_k : k < \omega \rangle), F(\xi)$ 54
$\dashv_P$ 59, 65
Rfn_S 62
RFN_S 63
$Prf_{S,\Gamma}(x,y), Pr_{S,\Gamma}(x)$ 63
$Con(n,S), Con_S^\omega$ 65
$Rfn(n,S), Rfn_S^\omega$ 65
$\dashv_{\Pi_1}$ 66
$Rfn_S(\Gamma), RFN_S(\Gamma)$ 69
RFN_τ 69
$[\Gamma]_S(x,y)$ 75
$\dashv_\Gamma$ 76
$[\Gamma](x,y)$ 76
$Cons(\Gamma,T)$ 80
$\dotdiv$ 82
$\Vdash$ 84
$rank, rank_n$ 84, 85
$\lambda(x)$ 84
p^I 84
$\{\Sigma_n\}(x,u,v,w)$ 89
$\chi_n(x,y,z)$ 93
$t, \mu_t(x)$ 96
$t: S' \leq S$ 97
$\leq, <$ 97

Hcm_ξ 100
$\equiv$ 105
$Int_{A,B}$ 106
$\triangleleft$ 106
$\simeq$ 109
$\leqslant, d(A), D_T, \mathbf{D}_T$ 118
$A^T, \downarrow, \uparrow, \vee$ 118
$\cap, \cup$ 119
$0_T, 1_T$ 119
$\ll_\cap, \ll_\cup$ 121
$a + \sigma$ 121
CON_T 122
$\cup, \cap$ 124
$[a,b], [a,b),$ etc.   125
$d(\varphi), d(X)$ 127
$\ll$ 127
$\sigma, \sigma_0, \ldots, \pi, \pi_0, \ldots$ 127
$\Phi(B_n, \Delta_n, \Gamma, \Pi_n, \Sigma_n)$ 127
$Cl^\cup(\Phi), Cl(\Phi)$ 132
$\lll$ 132, 133
$t^{-1}(T)$ 146
$\leqslant^r, \leqslant^{er}$ 147
L_{ST}, ZF 147
t_s 147
$\nu_n(x)$ 148
$H_k, H_k(x)$ 148
$Tr_k(x)$ 148
RFN_S^t 148
$\Sigma_k^{ST}, \Pi_k^{ST}$ 149
GB   152

LECTURE NOTES IN LOGIC

General Remarks

This series is intended to serve researchers, teachers, and students in the field of symbolic logic, broadly interpreted. The aim of the series is to bring publications to the logic community with the least possible delay and to provide rapid dissemination of the latest research. Scientific quality is the overriding criterion by which submissions are evaluated.

Books in the Lecture Notes in Logic series are printed by photo-offset from master copy prepared using LaTeX or (preferably) $\mathcal{AMS}$-LaTeX and the ASL style files. For this purpose the Association for Symbolic Logic provides technical instructions to authors. Careful preparation of manuscripts will help keep production time short, reduce costs, and ensure quality of appearance of the finished book. Authors receive 50 free copies of their book. No royalty is paid on LNL volumes.

Commitment to publish may be made by letter of intent rather than by signing a formal contract, at the discretion of the ASL Publisher. The Association for Symbolic Logic secures the copyright for each volume.

The editors prefer email contact and encourage electronic submissions.

Editorial Board

Samuel R. Buss, Managing Editor
Department of Mathematics
University of California, San Diego
La Jolla, California 92093-0112, USA
sbuss@ucsd.edu

Lance Fortnow
NEC Research Institute
4 Independence Way
Princeton, New Jersey 08540, USA
fortnow@research.nj.nec.com

Shaughan Lavine
Department of Philosophy
The University of Arizona
P.O. Box 210027
Tuscon, Arizona 85721-0027, USA
shaughan@ns.arizona.edu

Steffen Lempp
Department of Mathematics
University of Wisconsin
480 Lincoln Avenue
Madison, Wisconsin 53706-1388, USA
lempp@math.wisc.edu

Anand Pillay
Department of Mathematics
University of Illinois
1409 West Green Street
Urbana, Illinois 61801, USA
pillay@math.uiuc.edu

W. Hugh Woodin
Department of Mathematics
University of California, Berkeley
Berkeley, California 94720, USA
woodin@math.berkeley.edu

Editorial Policy

1. Submissions are invited in the following categories:
i) Research monographs
ii) Lecture and seminar notes
iii) Reports of meetings
iv) Texts which are out of print

Those considering a project which might be suitable for the series are strongly advised to contact the publisher or the series editors at an early stage.

2. Categories i) and ii). These categories will be emphasized by Lecture Notes in Logic and are normally reserved for works written by one or two authors. The goal is to report new developments quickly, informally, and in a way that will make them accessible to non-specialists. Books in these categories should include
– at least 100 pages of text;
– a table of contents;
– an informative introduction, perhaps with some historical remarks, which should be accessible to readers unfamiliar with the topic treated;
– a subject index.

In the evaluation of submissions, timeliness of the work is an important criterion. Texts should be well-rounded and reasonably self-contained. In most cases the work will contain results of others as well as those of the authors. In each case, the author(s) should provide sufficient motivation, examples, and applications. In this respect, Ph.D. theses will be suitable for this series only when they are of exceptional interest and of high expository quality.

Proposals for volumes in this category should be submitted (preferably in duplicate) to one of the series editors, and will be refereed. A provisional judgment on the acceptability of a project can be based on partial information about the work: a first draft, or a detailed outline describing the contents of each chapter, the estimated length, a bibliography, and one or two sample chapters. A final decision whether to accept will rest on an evaluation of the completed work.

3. Category iii). Reports of meetings will be considered for publication provided that they are of lasting interest. In exceptional cases, other multi-authored volumes may be considered in this category. One or more expert participant(s) will act as the scientific editor(s) of the volume. They select the papers which are suitable for inclusion and have them individually refereed as for a journal. Organizers should contact the Managing Editor of Lecture Notes in Logic in the early planning stages.

4. Category iv). This category provides an avenue whereby out-of-print books that are still in demand can be made available to a new generation of logicians.

5. Format. Works in English are preferred. After the manuscript is accepted in its final form, an electronic copy in LaTeX or (preferably) $\mathcal{AMS}$-LaTeX format will be appreciated and will advance considerably the publication date of the book. Authors are strongly urged to seek typesetting instructions from the Association for Symbolic Logic at an early stage of their manuscript preparation.

LECTURE NOTES IN LOGIC

From 1993 to 1999 this series was published under an agreement between the Association for Symbolic Logic and Springer-Verlag. Since 1999 the ASL is Publisher and A K Peters, Ltd. is Co-publisher. The ASL is committed to keeping all books in the series in print.

Current information may be found at http://www.aslonline.org, the ASL Web site. Editorial and submission policies and the list of Editors may also be found above.

Previously published books in the *Lecture Notes in Logic* are:

1. *Recursion theory.* J. R. Shoenfield. (1993, reprinted 2001; 84 pp.)

2. *Logic Colloquium '90; Proceedings of the Annual European Summer Meeting of the Association for Symbolic Logic, held in Helsinki, Finland, July 15–22, 1990.* Eds. J. Oikkonen and J. Väänänen. (1993, reprinted 2001; 305 pp.)

3. *Fine structure and iteration trees.* W. Mitchell and J. Steel. (1994; 130 pp.)

4. *Descriptive set theory and forcing: how to prove theorems about Borel sets the hard way.* A. W. Miller. (1995; 130 pp.)

5. *Model theory of fields.* D. Marker, M. Messmer, and A. Pillay. (1996; 154 pp.)

6. *Gödel '96; Logical foundations of mathematics, computer science and physics; Kurt Gödel's legacy. Brno, Czech Republic, August 1996, Proceedings.* Ed. P. Hajek. (1996, reprinted 2001; 322 pp.)

7. *A general algebraic semantics for sentential objects.* J. M. Font and R. Jansana. (1996; 135 pp.)

8. *The core model iterability problem.* J. Steel. (1997; 112 pp.)

9. *Bounded variable logics and counting.* M. Otto. (1997; 183 pp.)

10. *Aspects of incompleteness.* P. Lindstrom. (1997, 2nd edition 2003; 163 pp.)

11. *Logic Colloquium '95; Proceedings of the Annual European Summer Meeting of the Association for Symbolic Logic, held in Haifa, Israel, August 9–18, 1995.* Eds. J. A. Makowsky and E. V. Ravve. (1998; 364 pp.)

12. *Logic Colloquium '96; Proceedings of the Colloquium held in San Sebastian, Spain, July 9–15, 1996.* Eds. J. M. Larrazabal, D. Lascar, and G. Mints. (1998; 268 pp.)

13. *Logic Colloquium '98; Proceedings of the Annual European Summer Meeting of the Association for Symbolic Logic, held in Prague, Czech Republic, August 9–15, 1998.* Eds. S. R. Buss, P. Hájek, and P. Pudlák. (2000; 541 pp.)

14. *Model Theory of Stochastic Processes.* S. Fajardo and H. J. Keisler. (2002; 136 pp.)

15. *Reflections on the Foundations of Mathematics; Essays in honor of Solomon Feferman.* Eds. W. Seig, R. Sommer, and C. Talcott. (2002; 444 pp.)

16. *Inexhaustibility; a non-exhaustive treatment.* T. Franzén. (2003; 255 pp.)

17. *Logic Colloquium '99; Proceedings of the Annual European Summer Meeting of the Association for Symbolic Logic, held in Utrecht, Netherlands, August 1–6, 1999.* Eds. J. van Eijck, V. van Oostrom, and A. Visser. (2003; 208 pp.)